INVESTIGATIONS IN NUMBER, DATA, AND SPACE®

Exploring Pattern

Pattern Trains and Hopscotch Paths

Kindergarten

Rebeka Eston
Karen Economopoulos

Developed at TERC, Cambridge, Massachusetts

Dale Seymour Publications®
White Plains, New York

The *Investigations* curriculum was developed at TERC (formerly Technical Education Research Centers) in collaboration with Kent State University and the State University of New York at Buffalo. The work was supported in part by National Science Foundation Grant No. ESI-9050210. TERC is a nonprofit company working to improve mathematics and science education. TERC is located at 2067 Massachusetts Avenue, Cambridge, MA 02140.

This project was supported, in part,
by the
National Science Foundation
Opinions expressed are those of the authors
and not necessarily those of the Foundation

Managing Editor: Catherine Anderson
Series Editor: Beverly Cory
ESL Consultant: Nancy Sokol Green
Production/Manufacturing Director: Janet Yearian
Production/Manufacturing Manager: Karen Edmonds
Production/Manufacturing Coordinators: Joe Conte/Roxanne Knoll
Design Manager: Jeff Kelly
Design: Don Taka
Composition: Archetype Book Composition
Illustrations: Susan Jaekel, Rachel Gage, Carl Yoshihara
Cover: Bay Graphics

This book is published by Dale Seymour Publications®, an imprint of Addison Wesley Longman, Inc.

> Dale Seymour Publications
> 10 Bank Street
> White Plains, NY 10602
> Customer Service: 1-800-872-1100

Order number DS47104
ISBN 1-57232-927-0
3 4 5 6 7 8 9 10-ML-02 01 00 99

 Printed on Recycled Paper

T E R C

Principal Investigator   Susan Jo Russell

Co-Principal Investigator   Cornelia Tierney

Director of Research and Evaluation   Jan Mokros

Director of K–2 Curriculum   Karen Economopoulos

Curriculum Development
Karen Economopoulos
Rebeka Eston
Marlene Kliman
Christopher Mainhart
Jan Mokros
Megan Murray
Kim O'Neil
Susan Jo Russell
Tracey Wright

Evaluation and Assessment
Mary Berle-Carman
Jan Mokros
Andee Rubin

Teacher Support
Irene Baker
Megan Murray
Kim O'Neil
Judy Storeygard
Tracey Wright

Technology Development
Michael T. Battista
Douglas H. Clements
Julie Sarama

Video Production
David A. Smith
Judy Storeygard

Administration and Production
Irene Baker
Amy Catlin

**Cooperating Classrooms
for This Unit**
Jeanne Wall
Arlington Public Schools
Arlington, MA

Audrey Barzey
Patricia Kelliher
Ellen Tait
Boston Public Schools
Boston, MA

Meg Bruton
Fayerweather Street School
Cambridge, MA

Rebeka Eston
Lincoln Public Schools
Lincoln, MA

Lila Austin
The Atrium School
Watertown, MA

Christopher Mainhart
Westwood Public Schools
Westwood, MA

Consultants and Advisors
Deborah Lowenberg Ball
Michael T. Battista
Marilyn Burns
Douglas H. Clements
Ann Grady

CONTENTS

TEACHER NOTES

WHERE TO START

The first-time user of *Pattern Trains and Hopscotch Paths* should read the following:

When you next teach this same unit, you can begin to read more of the background. Each time you present the unit, you will learn more about how your students understand the mathematical ideas.

Investigations in Number, Data, and Space® is a K–5 mathematics curriculum with four major goals:

■ to offer students meaningful mathematical problems

■ to emphasize depth in mathematical thinking rather than superficial exposure to a series of fragmented topics

■ to communicate mathematics content and pedagogy to teachers

■ to substantially expand the pool of mathematically literate students

The *Investigations* curriculum embodies a new approach based on years of research about how children learn mathematics. Each grade level consists of a set of separate units, each offering 2–8 weeks of work. These units of study are presented through investigations that involve students in the exploration of major mathematical ideas.

Approaching the mathematics content through investigations helps students develop flexibility and confidence in approaching problems, fluency in using mathematical skills and tools to solve problems, and proficiency in evaluating their solutions. Students also build a repertoire of ways to communicate about their mathematical thinking, while their enjoyment and appreciation of mathematics grows.

The investigations are carefully designed to invite all students into mathematics—girls and boys, members of diverse cultural, ethnic, and language groups, and students with different strengths and interests. Problem contexts often call on students to share experiences from their family, culture, or community. The curriculum eliminates barriers—such as work in isolation from peers, or emphasis on speed and memorization—that exclude some students from participating successfully in mathematics. The following aspects of the curriculum ensure that all students are included in significant mathematics learning:

■ Students spend time exploring problems in depth.

■ They find more than one solution to many of the problems they work on.

■ They invent their own strategies and approaches, rather than rely on memorized procedures.

■ They choose from a variety of concrete materials and appropriate technology, including calculators, as a natural part of their everyday mathematical work.

■ They express their mathematical thinking through drawing, writing, and talking.

■ They work in a variety of groupings—as a whole class, individually, in pairs, and in small groups.

■ They move around the classroom as they explore the mathematics in their environment and talk with their peers.

While reading and other language activities are typically given a great deal of time and emphasis in elementary classrooms, mathematics often does not get the time it needs. If students are to experience mathematics in depth, they must have enough time to become engaged in real mathematical problems. We believe that a minimum of 5 hours of mathematics classroom time a week—about an hour a day—is critical at the elementary level. The scope and pacing of the *Investigations* curriculum are based on that belief.

We explain more about the pedagogy and principles that underlie these investigations in Teacher Notes throughout the units. For correlations of the curriculum to the NCTM Standards and further help in using this research-based program for teaching mathematics, see the following books, available from Dale Seymour Publications:

■ *Implementing the* Investigations in Number, Data, and Space® *Curriculum*

■ *Beyond Arithmetic: Changing Mathematics in the Elementary Classroom* by Jan Mokros, Susan Jo Russell, and Karen Economopoulos

This book is one of the curriculum units for *Investigations in Number, Data, and Space.* In addition to providing part of a complete mathematics curriculum for your students, this unit offers information to support your own professional development. You, the teacher, are the person who will make this curriculum come alive in the classroom; the book for each unit is your main support system.

Although the curriculum does not include student instructional texts, reproducible sheets for student work are provided with the units and, in some cases, are also available as Student Activity Booklets. In these investigations, students work actively with objects and experiences in their own environment, including manipulative materials and technology, rather than with a workbook.

Ultimately, every teacher will use these investigations in ways that make sense for his or her particular style, the particular group of students, and the constraints and supports of a particular school environment. Each unit offers information and guidance drawn from our collaborations with many teachers and students over many years. Our goal is to help you, a professional educator, give all your students access to mathematical power.

Investigation Format

The opening two pages of each investigation help you get ready for the work that follows.

- **Focus Time** This gives a synopsis of the activities used to introduce the important mathematical ideas for the investigation.

- **Choice Time** This lists the activities, new and recurring, that support the Focus Time work.

- **Mathematical Emphasis** This highlights the most important ideas and processes students will encounter in this investigation.

- **Teacher Support** This indicates the Teacher Notes and Dialogue Boxes included to help you understand what's going on mathematically in your classroom.

- **What to Plan Ahead of Time** These lists alert you to materials to gather, sheets to duplicate, and other things you need to do before starting the investigation. Full details of materials and preparation are included with each activity.

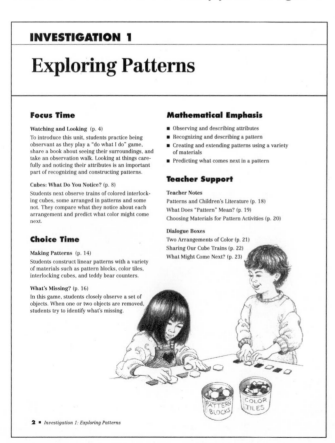

INVESTIGATION 1

Exploring Patterns

Focus Time

Watching and Looking (p. 4)
To introduce this unit, students practice being observant as they play a "do what I do" game, share a book about seeing their surroundings, and take an observation walk. Looking at things carefully and noticing their attributes is an important part of recognizing and constructing patterns.

Cubes: What Do You Notice? (p. 8)
Students next observe trains of colored interlocking cubes, some arranged in patterns and some not. They compare what they notice about each arrangement and predict what color might come next.

Choice Time

Making Patterns (p. 14)
Students construct linear patterns with a variety of materials such as pattern blocks, color tiles, interlocking cubes, and teddy bear counters.

What's Missing? (p. 16)
In this game, students closely observe a set of objects. When one or two objects are removed, students try to identify what's missing.

Mathematical Emphasis

- Observing and describing attributes
- Recognizing and describing a pattern
- Creating and extending patterns using a variety of materials
- Predicting what comes next in a pattern

Teacher Support

Teacher Notes
Patterns and Children's Literature (p. 18)
What Does "Pattern" Mean? (p. 19)
Choosing Materials for Pattern Activities (p. 20)

Dialogue Boxes
Two Arrangements of Color (p. 21)
Sharing Our Cube Trains (p. 22)
What Might Come Next? (p. 23)

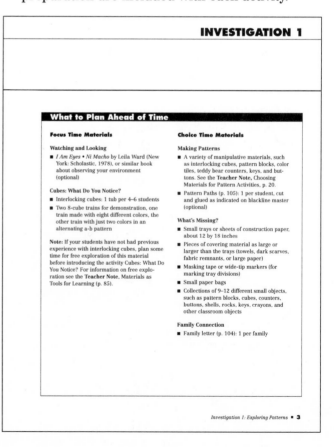

INVESTIGATION 1

What to Plan Ahead of Time

Focus Time Materials

Watching and Looking
- *I Am Eyes • Ni Macho* by Leila Ward (New York: Scholastic, 1978), or similar book about observing your environment (optional)

Cubes: What Do You Notice?
- Interlocking cubes: 1 tub per 4–6 students
- Two 8-cube trains for demonstration, one train made with eight different colors, the other train with just two colors in an alternating a-b pattern

Note: If your students have not had previous experience with interlocking cubes, plan some time for free exploration of this material before introducing the activity Cubes: What Do You Notice? For information on free exploration see the **Teacher Note**, Materials as Tools for Learning (p. 85).

Choice Time Materials

Making Patterns
- A variety of manipulative materials, such as interlocking cubes, pattern blocks, color tiles, teddy bear counters, keys, and buttons. See the Teacher Note, Choosing Materials for Pattern Activities, p. 20.
- Pattern Paths (p. 105): 1 per student, cut and glued as indicated on blackline master (optional)

What's Missing?
- Small trays or sheets of construction paper, about 12 by 18 inches
- Pieces of covering material as large or larger than the trays (towels, dark scarves, fabric remnants, or large paper)
- Masking tape or wide-tip markers (for marking tray divisions)
- Small paper bags
- Collections of 9–12 different small objects, such as pattern blocks, cubes, counters, buttons, shells, rocks, keys, crayons, and other classroom objects

Family Connection
- Family letter (p. 104): 1 per family

Always read through an entire investigation before you begin, in order to understand the overall flow and sequence of the activities.

Focus Time In this whole-group meeting, you introduce one or more activities that embody the important mathematical ideas underlying the investigation. The group then may break up into individuals or pairs for further work on the same activity. Many Focus Time activities culminate with a brief sharing time or discussion as a way of acknowledging students' work and highlighting the mathematical ideas. Focus Time varies in length. Sometimes it is short and can be completed in a single group meeting or a single work period; other times it may stretch over two or three sessions.

Choice Time Each Focus Time is followed by Choice Time, which offers a series of supporting activities to be done simultaneously by individuals, pairs, or small groups. You introduce these related tasks over a period of several days. During Choice Time, students work independently, at their own pace, choosing the activities they prefer and often returning many times to their favorites. Many kindergarten classrooms have an activity time built into their daily schedule, and Choice Time activities can easily be incorporated.

Together, the Focus Time and Choice Time activities offer a balanced kindergarten curriculum.

Classroom Routines The kindergarten day is filled with opportunities to work with mathematics. Routines such as taking attendance, asking about snack preferences, and discussing the calendar offer regular, ongoing practice in counting, collecting and organizing data, and understanding time.

Four specific routines—Attendance, Counting Jar, Calendar, and Today's Question—are formally introduced in the unit *Mathematical Thinking in Kindergarten*. Another routine, Patterns on the Pocket Chart, is introduced in the unit *Pattern Trains and Hopscotch Paths*. Descriptions of these routines can also be found in an appendix for each unit, and reminders of their ongoing use appear in the Unit Overview charts.

The Linguistically Diverse Classroom Each unit includes an appendix with Tips for the Linguistically Diverse Classroom to help teachers support students at varying levels of English proficiency. While more specific tips appear within the units at grades 1–5, often in relation to written work, general tips on oral discussions and observing the students are more appropriate for kindergarten.

Also included are suggestions for vocabulary work to help ensure that students' linguistic difficulties do not interfere with their comprehension of math concepts. The Preview for the Linguistically Diverse Classroom lists key words in the unit that are generally known to English-speaking kindergartners. Activities to help familiarize other students with these words are found in the appendix, Vocabulary Support for Second-Language Learners. In addition, ideas for making connections to students' languages and cultures, included on the Preview page, help the class explore the unit's concepts from a multicultural perspective.

Materials

A complete list of the materials needed for teaching this unit follows the Unit Overview. These materials are available in *Investigations* kits or can be purchased from school supply dealers.

Classroom Materials In an active kindergarten mathematics classroom, certain basic materials should be available at all times, including interlocking cubes, a variety of things to count with, and writing and drawing materials. Some activities in this curriculum require scissors and glue sticks or tape; dot stickers and large paper are also useful. So that students can independently get what they need at any time, they should know where the materials are kept, how they are stored, and how they are to be returned to the storage area.

Children's Literature Each unit offers a list of children's literature that can be used to support the mathematical ideas in the unit. Sometimes an activity incorporates a specific children's book, with suggestions for substitutions where practical. While such activities can be adapted and taught without the book, the literature offers a rich introduction and should be used whenever possible. If you can get the titles in Big Book format, these are ideal for kindergarten.

Blackline Masters Student recording sheets and other teaching tools for both class and homework are provided as reproducible blackline masters at the end of each unit. When student sheets are designated for kindergarten homework, they usually repeat an activity from class, such as playing a game, as a way of involving and informing family members. Occasionally a homework sheet may ask students to collect data or materials for a class project or in preparation for upcoming activities.

Student Activity Booklets For the two kindergarten number units, the blackline masters are also available as Student Activity Booklets, designed to free you from extensive copying. The other kindergarten units require minimal copying.

Family Letter A letter that you can send home to students' families is included with the blackline masters for each unit. Families need to be informed about the mathematics work in your classroom; they should be encouraged to participate in and support their children's work. A reminder to send home the letter for each unit appears in one of the early investigations. These letters are also available separately in Spanish, Vietnamese, Cantonese, Hmong, and Cambodian.

Investigations **at Home** To further involve families in the kindergarten program, you can offer them the *Investigations* at Home booklet, which describes the kindergarten units, explains the mathematics work children do in kindergarten, and offers activities families can do with their children at home.

Adapting *Investigations* to Your Classroom

Kindergarten programs vary greatly in the amount of time each day that students attend. We recommend that kindergarten teachers devote from 30 to 45 minutes daily to work in mathematics, but we recognize that this can be challenging in a half-day program. The kindergarten level of *Investigations* is intentionally flexible so that teachers can adapt the curriculum to their particular setup.

Kindergartens participating in the *Investigations* field test included full-day programs, half-day programs of approximately 3 hours, and half-day programs that add one or two full days to the kindergarten week at some point in the school year. Despite the wide range of program structures, classrooms generally fell into one of two groups: those that offered a separate math time daily (Math Workshop or Math Time), and those that included one or two mathematics activities during a general Activity Time or Station Time.

Math Workshop Teachers using a Math Workshop approach set aside 30 to 45 minutes each day for doing mathematics. In addition, they usually also have a more general activity time in their daily schedule. On some days, Math Workshop might be devoted to the Focus Time activities, with the whole class gathered together. On other days, students might work in small groups and choose from three or four Choice Time activities.

Math as Part of Activity Time Teachers with less time in their day may offer students one or two math activities, along with activities from other areas of the curriculum, during their Activity Time or Station Time. For example, on a particular day, students might be able to choose among a science activity, block building, an art project, dramatic play, books, puzzles, and a math activity. New activities are introduced during a whole-class meeting. With the *Investigations* curriculum, teachers who use this approach have found that it is important to designate at least one longer block of time (30 to 45 minutes) each week for mathematics. During this time, students engage in Focus Time activities and have a chance to share their work and discuss mathematical ideas. The suggested Choice Time activities are then presented as part of the general activity time. Following this model, work on a curriculum unit will naturally stretch over a longer period.

Planning Your Curriculum The amount of time scheduled for mathematics work will determine how much of the kindergarten *Investigations* curriculum a teacher is able to cover in the school year. You may have to make some choices as you adapt the units to your particular schedule. What is most important is finding a way to involve students in mathematics every day of the school year.

Each unit will be handled somewhat differently by every teacher. You need to be active in determining an appropriate pace and the best transition points for your class. As you read an investigation, make some preliminary decisions about how many days you will need to present the activities, based on what you know about your students and about your schedule. You may need to modify your initial plans as you proceed, and you may want to make notes in the margins of the pages as reminders for the next time you use the unit.

Help for You, the Teacher

Because we believe strongly that a new curriculum must help teachers think in new ways about mathematics and about their students' mathematical thinking processes, we have included a great deal of material to help you learn more about both.

About the Mathematics in This Unit This introductory section summarizes the essential information about the mathematics you will be teaching. It describes the unit's central mathematical ideas and the ways students will encounter them through the unit's activities.

Teacher Notes These reference notes provide practical information about the mathematics you are teaching and about our experience with how students learn. Many of the notes were written in response to actual questions from teachers or to discuss important things we saw happening in the field-test classrooms. Some teachers like to read them all before starting the unit, then review them as they come up in particular investigations.

In the kindergarten units, Teacher Notes headed "From the Classroom" contain anecdotal reflections of teachers. Some focus on classroom management issues, while others are observations of students at work. These notes offer another perspective on how an activity might unfold or how kindergarten students might become engaged with a particular material or activity.

A few Teacher Notes touch on fundamental principles of using *Investigations* and focus on the pedagogy of the kindergarten classroom:

- About Choice Time
- Materials as Tools for Learning
- Encouraging Students to Think, Reason, and Share Ideas
- Games: The Importance of Playing More Than Once

After their initial appearance, these are repeated in the back of each unit. Reviewing these notes periodically can help you reflect on important aspects of the *Investigations* curriculum.

Dialogue Boxes Sample dialogues demonstrate how students typically express their mathematical ideas, what issues and confusions arise in their thinking, and how some teachers have guided class discussions.

Many of these dialogues are word-for-word transcriptions of recorded class discussions. They are not always easy reading; sometimes it may take some effort to unravel what the students are trying to say. But this is the value of these dialogues; they offer good clues to how your students may develop and express their approaches and strategies, helping you prepare for your own class discussions.

Where to Start You may not have time to read everything the first time you use this unit. As a first-time user, you will likely focus on understanding the activities and working them out with your students. You will also want to read the few sections listed in the Contents under the heading Where to Start.

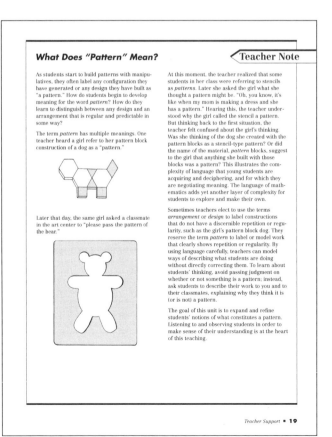

The *Investigations* curriculum incorporates the use of two forms of technology in the classroom: calculators and computers. Calculators are assumed to be standard classroom materials, available for student use in any unit. Computers are explicitly linked to one or more units at each grade level; they are used with the unit on 2-D geometry at each grade, as well as with some of the units on measuring, data, and changes.

Using Calculators

In this curriculum, calculators are considered tools for doing mathematics, similar to pattern blocks or interlocking cubes. Just as with other tools, students must learn both *how* to use calculators correctly and *when* they are appropriate to use. This knowledge is crucial for daily life, as calculators are now a standard way of handling numerical operations, both at work and at home. Calculators are formally introduced in the grade 1 curriculum, but if available, can be introduced to kindergartners informally.

Using a calculator correctly is not a simple task; it depends on a good knowledge of the four operations and of the number system, so that students can select suitable calculations and also determine what a reasonable result would be. These skills are the basis of any work with numbers, whether or not a calculator is involved.

Unfortunately, calculators are often seen as tools to check computations with, as if other methods are somehow more fallible. Students need to understand that any computational method can be used to check any other; it's just as easy to make a mistake on the calculator as it is to make a mistake on paper or with mental arithmetic. Throughout this curriculum, we encourage students to solve computation problems in more than one way in order to double-check their accuracy. We present mental arithmetic, paper-and-pencil computation, and calculators as three possible approaches.

In this curriculum we also recognize that, despite their importance, calculators are not always appropriate in mathematics instruction. Like any tools, calculators are useful for some tasks but not for others. You will need to make decisions about when to allow students access to calculators and when to ask that they solve problems without them so that they can concentrate on other tools and skills. At times when calculators are or are not appropriate for a particular activity, we make specific recommendations. Help your students develop their own sense of which problems they can tackle with their own reasoning and which ones might be better solved with a combination of their own reasoning and the calculator.

Managing calculators in your classroom so that they are a tool, and not a distraction, requires some planning. When calculators are first introduced, students often want to use them for everything, even problems that can be solved quite simply by other methods. However, once the novelty wears off, students are just as interested in developing their own strategies, especially when these strategies are emphasized and valued in the classroom. Over time, students will come to recognize the ease and value of solving problems mentally, with paper and pencil, or with manipulatives, while also understanding the power of the calculator to facilitate work with larger numbers.

Experience shows that if calculators are available only occasionally, students become excited and distracted when permitted to use them. They focus on the tool rather than on the mathematics. In order to learn when calculators are appropriate and when they are not, students must have easy access to them and use them routinely in their work.

If you have a calculator for each student, and if you think your students can accept the responsibility, you might allow them to keep their calculators with the rest of their individual materials, at least for the first few weeks of school. Alternatively, you might store them in boxes on a shelf, number each calculator, and assign a corresponding number to each student. This system can give students a sense of ownership while also helping you keep track of the calculators.

Using Computers

Students can use computers to approach and visualize mathematical situations in new ways. The computer allows students to construct and manipulate geometric shapes, see objects move according to rules they specify, and turn, flip, and repeat a pattern.

This curriculum calls for computers in units where they are a particularly effective tool for learning mathematics content. One unit on 2-D geometry at each of the grades 3–5 includes a core of activities that rely on access to computers, either in the classroom or in a lab. Other units on geometry, measuring, data, and changes include computer activities, but can be taught without them. In these units, however, students' experience is greatly enhanced by computer use.

The following list outlines the recommended use of computers in this curriculum:

Kindergarten
Unit: *Making Shapes and Building Blocks*
    (Exploring Geometry)
Software: *Shapes*
Source: provided with the unit

Grade 1
Unit: *Survey Questions and Secret Rules*
    (Collecting and Sorting Data)
Software: *Tabletop, Jr.*
Source: Broderbund

Unit: *Quilt Squares and Block Towns*
    (2-D and 3-D Geometry)
Software: *Shapes*
Source: provided with the unit

Grade 2
Unit: *Mathematical Thinking at Grade 2*
    (Introduction)
Software: *Shapes*
Source: provided with the unit

Unit: *Shapes, Halves, and Symmetry*
    (Geometry and Fractions)
Software: *Shapes*
Source: provided with the unit

Unit: *How Long? How Far?* (Measuring)
Software: *Geo-Logo*
Source: provided with the unit

Grade 3
Unit: *Flips, Turns, and Area* (2-D Geometry)
Software: *Tumbling Tetrominoes*
Source: provided with the unit

Unit: *Turtle Paths* (2-D Geometry)
Software: *Geo-Logo*
Source: provided with the unit

Grade 4
Unit: *Sunken Ships and Grid Patterns*
    (2-D Geometry)
Software: *Geo-Logo*
Source: provided with the unit

Grade 5
Unit: *Picturing Polygons* (2-D Geometry)
Software: *Geo-Logo*
Source: provided with the unit

Unit: *Patterns of Change* (Tables and Graphs)
Software: *Trips*
Source: provided with the unit

Unit: *Data: Kids, Cats, and Ads* (Statistics)
Software: *Tabletop, Sr.*
Source: Broderbund

The software provided with the *Investigations* units uses the power of the computer to help students explore mathematical ideas and relationships that cannot be explored in the same way with physical materials. With the *Shapes* (grades K–2) and *Tumbling Tetrominoes* (grade 3) software, students explore symmetry, pattern, rotation and reflection, area, and characteristics of 2-D shapes. With the *Geo-Logo* software (grades 2–5), students investigate rotations and reflections, coordinate geometry, the properties of 2-D shapes, and angles. The *Trips* software (grade 5) is a mathematical exploration of motion in which students run experiments and interpret data presented in graphs and tables.

We suggest that students work in pairs on the computer; this not only maximizes computer resources but also encourages students to consult, monitor, and teach one another. However, asking more than two students to work at the same computer is less effective. Managing access to computers is an issue for every classroom. The curriculum gives you explicit support for setting up a system. The units are structured on the assumption that you have enough computers for half your students to work on the machines in pairs at one time. If you do not have access to that many computers, suggestions are made for structuring class time to use the unit with fewer than five.

Assessment plays a critical role in teaching and learning, and it is an integral part of the *Investigations* curriculum. For a teacher using these units, assessment is an ongoing process. You observe students' discussions and explanations of their ideas and strategies on a daily basis and examine their work as it evolves. While students are busy working with materials, playing mathematical games, sharing ideas with partners, and working on projects, you have many opportunities to observe their mathematical thinking. What you learn through observation guides your decisions about how to proceed, both with the curriculum and with individual students.

Our experiences with young children suggest that they know, can explain, and can demonstrate with materials a lot more than they can represent on paper. This is one reason why it is so important to engage children in conversation, helping them explain their thinking about a problem they are solving. It is also why, in kindergarten, assessment is based exclusively on a teacher's observations of students as they work.

The way you observe students will vary throughout the year. At times you may be interested in particular strategies that students are developing to solve problems. Other times, you might want to observe how students use or do not use materials for solving problems. You may want to focus on how students interact when working in pairs or groups. You may be interested in noting the strategy that a student uses when playing a game during Choice Time. Or you may take note of student ideas and thinking during class discussions.

Assessment Tools in the Unit

Virtually every activity in the kindergarten units of the *Investigations* curriculum includes a section called Observing the Students. This section is a teacher's primary assessment tool. It offers guidelines on what to look for as students encounter the mathematics of the activity. It may suggest questions you can ask to uncover student thinking or to stimulate further investigation. When useful, a range of potential responses or examples of typical student approaches is given, along with ways to adapt the activity for students in need of more or less challenge.

Supplementing this main assessment tool in each unit are the Teacher Notes and Dialogue Boxes that contain examples of student work, teacher observations, and student conversations from real kindergarten classrooms. These resources can help you interpret experiences from your own classroom as you progress through a unit.

Documentation of Student Growth

You will probably need to develop some sort of system to record and keep track of your observations. A single observation is like a snapshot of a student's experience with a particular activity, but when considered over time, a collection of these snapshots provides an informative and detailed picture of a student. Such observations are useful in documenting and assessing students' growth, as well as in planning curriculum.

Observation Notes A few ideas that teachers have found successful for record keeping are suggested here. The most important consideration is finding a system that really works for you. All too often, keeping observation notes on a class of 20–30 students is overwhelming and time-consuming. Your goal is to find a system that is neither.

Some teachers find that a class list of names is convenient for jotting down their observations. Since the space is limited, it is not possible to write lengthy notes; however, over time, these short observations provide important information.

Other teachers keep a card file or a loose-leaf notebook with a page for each student. When something about a student's thinking strikes them as important, they jot down brief notes and the date.

Some teachers use self-sticking address labels, kept on clipboards around the classroom. After taking notes on individual students, they simply peel off each label and stick it into the appropriate student file or notebook page.

You may find that writing notes at the end of each week works well for you. For some teachers, this process helps them reflect on individual students, on the curriculum, and on the class as a whole. Planning for the next weeks' activities often grows out of these weekly reflections.

Student Portfolios Collecting samples of student work from each unit in a portfolio is another way to document a student's experience that supports your observation notes. In kindergarten, samples of student work may include constructions, patterns, or designs that students have recorded, score sheets from games they have played, and early attempts to record their problem-solving strategies on paper, using pictures, numbers, or words.

The ability to record and represent one's ideas and strategies on paper develops over time. Not all 5- and 6-year-olds will be ready for this. Even when students are ready, what they record will have meaning for them only in the moment—as they work on the activity and make their representation. You can augment this by taking dictation of a student's idea or strategy. This not only helps both you and the student recall the idea, but also gives students a model of how their ideas could be recorded on paper.

Over the school year, student work samples combined with anecdotal observations are valuable resources when you are preparing for family conferences or writing student reports. They help you communicate student growth and progress, both to families and to the students' subsequent teachers.

Assessment Overview

There are two places to turn for a preview of the assessment information in each kindergarten *Investigations* unit. The Assessment Resources column in the Unit Overview chart locates the Observing the Students section for each activity, plus any Teacher Notes and Dialogue Boxes that explain what to look for and what types of responses you might see in your classroom. Additionally, the section called About the Assessment in This Unit gives you a detailed list of questions, keyed to the mathematical emphases for each investigation, to help you observe and assess student growth. This section also includes suggestions for choosing student work to save from the unit.

These examples illustrate record keeping systems used by two different teachers for the kindergarten unit *Collecting, Counting, and Measuring,* one using the class list and the other using individual note cards to record student progress.

Emma Ruiz

3/19 Counting Jar: counts 9 balls accurately and makes another set of 9 cubes

3/24 Today's Question: compares data, "13 is 4 more than 9 because the 13 tower is 4 names taller."

4/1 Draws Counting Book pictures for 1-6, then adds pgs 7, 8, 9, 10, 11 on her own

---

Unit: Collecting, Counting, and Measuring
Activity: Inventory Bags
Date: 10/12 and 10/13

Alexa • counting sequence to 50↑ • counts 1:1 up to 12 • counts 4 bags accurately	Luke • counts to 30, misses 19, 20 and 29, 30 • counts by moving objects; 1:1 to 10 objects • draws circles for buttons
Ayesha • works with Oscar • counts to 15 accurately – trouble beyond 15 but Oscar helps ★ meet to check • 1:1 to 8 objects? counting	Maddy • difficult to tell how much M. counted herself + how much was done by partner. Work w/ her to see.
Brendan absent 10/12, 10/13	Miyuki • counts aloud beyond 30 but leaves out 14 • counts 1:1 up to 10 but doesn't organize objects
Carlo • counts objects with difficulty. • remove items from bag so he works with 10 • says numbers to 10, counts objects to 6	Oscar • works with Ayesha • counts rotely to 20, maybe higher • double-checks his count every time – is accurate
Charlotte • completed inventory task easily without help • counts accurately up to 20 objects • represents with numbers	Ravi • worked w/ his aide to complete task • counts 1:1 to 5 objects • difficulty representing quantity w/ pictures
Felipe • worked well with Tarik • counted ~~~~~ bag – 21 in all	Renata

Pattern Trains and Hopscotch Paths

Content of This Unit Students investigate what makes a pattern. They look at the relationships among the parts of a pattern and focus on the information that allows us to predict what will come next. Using both math materials (color tiles, pattern blocks, interlocking cubes) and common objects (buttons, shells, keys), students copy, create, and extend simple linear patterns. In addition they explore how patterns are composed of the same repeating units.

In the second half of the unit, students extend their work with pattern as they make pattern paths similar to those used in the playground game hopscotch and investigate what happens when a linear pattern forms a rectangular border. They also explore patterns that grow or shrink in a rectangular and predictable way as they make "staircase" patterns with tiles and cubes.

Pattern Trains and Hopscotch Paths may be presented as a single unit, or it can be successfully divided into two parts to extend the work with pattern throughout the school year. For the latter option, teachers present Investigations 1 and 2 in the early part of kindergarten and return to Investigations 3 and 4 toward the end of the school year.

Connections with Other Units If you are doing the full-year *Investigations* curriculum in the suggested sequence for kindergarten, this is the second of six units. The work in looking for and identifying patterns and relationships lays a foundation for much of students' future work in mathematics in kindergarten and throughout elementary school.

Investigations Curriculum ■ Suggested Kindergarten Sequence

Mathematical Thinking in Kindergarten (Introduction)

▶ *Pattern Trains and Hopscotch Paths* (Exploring Pattern)

Collecting, Counting, and Measuring (Developing Number Sense)

Counting Ourselves and Others (Exploring Data)

Making Shapes and Building Blocks (Exploring Geometry)

How Many in All? (Counting and the Number System)

Investigation 1 ▪ Exploring Patterns

Class Sessions	Activities	Pacing
FOCUS TIME (p. 4) Watching and Looking	Can You Do What I Do? A Book About Looking Observation Walk	1–2 sessions
FOCUS TIME (p. 8) Cubes: What Do You Notice?	Two Arrangements of Cubes Making Cube Trains What Might Come Next? Homework: Family Connection	1–2 sessions
CHOICE TIME (p. 14)	Making Patterns What's Missing?	2–3 sessions
Classroom Routines	Attendance and Calendar (daily) Counting Jar, Today's Question, and Patterns on the Pocket Chart (weekly or as appropriate)	

Mathematical Emphasis

- Observing and describing attributes
- Recognizing and describing a pattern
- Creating and extending patterns
- Predicting what comes next in a pattern

Assessment Resources

Observing the Students:

- Two Arrangements of Cubes (p. 9)
- Making Cube Trains (p. 10)
- Making Patterns (p. 15)
- What's Missing? (p. 17)

Dialogue Box: Two Arrangements of Color (p. 21)

Dialogue Box: What Might Come Next (p. 23)

Materials

I Am Eyes • Ni Macho (optional)
Interlocking cubes
Small trays
Fabric or paper to conceal trays
Masking tape or markers
Paper bags
Small classroom objects
Pattern blocks
Color tiles
Teddy bear counters
Collections (e.g., keys, buttons)
Teaching resource sheets

Interlocking cubes

Investigation 2 ▪ What Comes Next?

Class Sessions	Activities	Pacing
FOCUS TIME (p. 26) Patterns on the Pocket Chart	Patterns on the Pocket Chart Patterns with Color Tiles	1–2 sessions
CHOICE TIME (p. 32)	What Comes Next? Pattern Block Snakes Add On Break the Train Make a Train	6–7 sessions
Classroom Routines Attendance and Calendar (daily) Counting Jar, Today's Question, and Patterns on the Pocket Chart (weekly or as appropriate)		

Mathematical Emphasis

- Recognizing a pattern
- Constructing and extending a pattern
- Reading a pattern
- Recording a pattern
- Predicting what comes next in a pattern
- Identifying the unit of a pattern

Assessment Resources

Observing the Students:

- Patterns with Color Tiles (p. 29)
- What Comes Next? (p. 33)
- Pattern Block Snakes (p. 35)
- Add On (p. 37)
- Break the Train (p. 39)
- Make a Train (p. 41)

Teacher Note: Reading Patterns (p. 43)

Dialogue Box: A "Harder" Pattern (p. 44)

Teacher Note: What's the Unit? (p. 45)

Materials

Hundred Number Wall Chart (for use as pocket chart)

Colored construction paper

Color tiles

Small paper cups

Resealable plastic bags

Paper strips

Glue sticks or paste

The Sultan's Snakes (optional)

Pattern blocks

Paper pattern blocks

Interlocking cubes

Blank 1-inch cubes

Stick-on labels

Teaching resource sheets

Pattern blocks

Investigation 3 ■ Hopscotch Paths

Class Sessions	Activities	Pacing
FOCUS TIME (p. 48) Hopscotch and Tile Paths	Jumping on a Hopscotch Path Making Hopscotch Paths Sharing Our Hopscotch Paths Paper Hopscotch Paths Extension: Hopscotch Around the World Extension: Pattern Museum Homework: Patterns from Home	2–3 sessions
CHOICE TIME (p. 56)	Hopscotch Paths Tile Paths Pattern Block Snakes Add On Break the Train Make a Train	5–6 sessions
Classroom Routines	Attendance and Calendar (daily) Counting Jar, Today's Question, and Patterns on the Pocket Chart (weekly or as appropriate)	

Mathematical Emphasis

- Constructing and extending a pattern
- Interpreting a pattern using physical movements
- Recording a pattern
- Representing a physical pattern using materials
- Predicting what comes next in a pattern
- Identifying the unit of a pattern

Assessment Resources

Observing the Students:

- Hopscotch Paths (pp. 51 and 57)
- Paper Hopscotch Paths (p. 53)
- Tile Paths (p. 59)

Teacher Note: What to Expect with Hopscotch Paths (p. 60)

Materials

Hopscotch Squares

Color tiles

Colored construction paper

Adding machine tape

Resealable plastic bags

Glue sticks or paste

Pattern blocks

Paper pattern blocks

Number cubes

Interlocking cubes

Teaching resource sheets

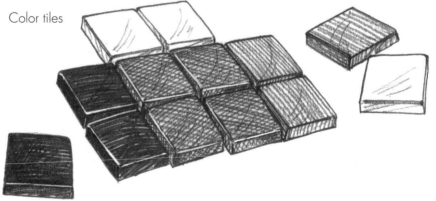

Color tiles

Investigation 4 ▪ Pattern Borders

Class Sessions	Activities	Pacing
FOCUS TIME (p. 66) Pattern Borders	A Border on the Pocket Chart Pattern Borders with Color Tiles Extension: Changing Borders on the Pocket Chart Extension: Students at the Pocket Chart	1–2 sessions
CHOICE TIME (p. 74)	Color Tile Borders 12 Chips Staircase Patterns Hopscotch Paths Tile Paths	4–5 sessions
Classroom Routines	Attendance and Calendar (daily) Counting Jar, Today's Question, and Patterns on the Pocket Chart (weekly or as appropriate)	

Mathematical Emphasis

- Making a linear pattern in a rectangular frame

- Making and comparing patterns that use the same two variables (of color)

- Copying, building, and extending patterns that grow (or shrink) in some regular and predictable way

- Determining a rule for how a pattern grows (or shrinks)

- Recording patterns

Assessment Resources

Observing the Students:

- Pattern Borders with Color Tiles (p. 71)

- Color Tile Borders (p. 75)

- 12 Chips (p. 77)

- Staircase Patterns (p. 79)

Dialogue Box: What Comes Here? (p. 80)

Materials

Hundred Number Wall Chart (for use as pocket chart)

Colored construction paper

Color tiles

Two-color counters

Colored pencils, markers, or crayons

Interlocking cubes

Hopscotch Squares

Adding machine tape

Resealable plastic bags

Glue sticks, tape

Student Sheets 1–4

Teaching resource sheets

Pocket chart

Following are the basic materials needed for the activities in this unit. Many items can be purchased from the publisher, either individually or in the Teacher Resource Package and the Student Materials Kit for kindergarten. Detailed information is available on the *Investigations* order form. To obtain this form, call toll-free 1-800-872-1100 and ask for a Dale Seymour customer service representative.

Snap™ cubes (interlocking cubes): class set, or 1 tub of 100 per 4–6 students

Pattern blocks: 1 bucket per 4–6 students

Paper pattern blocks: 2–3 sets

Color tiles: 2 sets of 400 per class

Teddy bear counters: 1 set per 4–6 students

Collections of items for making patterns, such as keys, buttons, shells, lids, bottle caps

Hundred Number Wall Chart, for use as pocket chart

I Am Eyes • Ni Macho by Leila Ward (optional)

The Sultan's Snakes by Lorna Turpin (optional)

Two-color counters: 12 per student

Blank 1-inch cubes, with stick-on labels for making number cubes and color cubes

Card stock and non-skid rug pad to make 8-inch Hopscotch Squares: 10–12 squares per pair or small group

Construction paper in assorted colors

Adding machine tape

Small paper cups: 2–3 dozen

Small trays or 12-by-18-inch sheets of construction paper: 3–4 for the class

Fabric or paper material to conceal each small tray

Masking tape

Small paper bags

Resealable plastic bags

Colored pencils, markers, crayons

Scissors

Glue sticks or paste

Assorted art supplies: rubber stamps and stamp pads, stencils, stickers, dot stickers, magazines and catalogs

The following materials are provided at the end of this unit as blackline masters.

Family Letter (p. 104)

Student Sheets 1–4 (p. 109)

Teaching Resources:

  Pattern Paths (p. 105)

  What Comes Next? Cards (p. 106)

  Color Cars (p. 107)

  Add to Our Pattern Museum (p. 108)

  Staircase Cards A–C (p. 113)

  Staircase Grid (p. 116)

  One-Inch Grid Paper (p. 117)

Related Children's Literature

Books About Observing

Martin, Bill, Jr. *Brown Bear, Brown Bear, What Do You See?* New York: Holt, 1992.

Ward, Leila. *I Am Eyes • Ni Macho.* New York: Scholastic, 1978.

Williams, Sue. *I Went Walking.* New York: Harcourt Brace Jovanovich, 1990.

Books About Hopscotch

Lankford, Mary. *Hopscotch Around the World.* New York: Morrow Junior Books, 1992.

Books with Pattern Borders

Chocolate, Debbi. *Kente Colors.* New York: Walker, 1996.

Dodds, Dayle Ann. *The Shape of Things.* Cambridge, MA: Candlewick Press, 1994.

Martin, B. & Archambault, J. *Chicka Chicka Boom Boom.* New York: Simon & Schuster, 1989.

Books with Illustrations Using Patterns

Angelou, Maya. *My Painted House, My Friendly Chicken and Me.* New York: Clarkson Potter, 1994.

Ganly, Helen. *Jyoti's Journey.* London: Andre Deutsch, 1986.

Grossman, Virginia. *Ten Little Rabbits.* San Francisco, CA: Chronicle Books, 1991.

Turpin, Lorna. *The Sultan's Snakes.* Swindon, England: Child's Play, 1996.

Students begin to appreciate the beauty and complexity of pattern as they identify patterns in the world around them, hear them in language and music, feel them in beats and motions, construct simple models and representations, and describe what they see and hear. Patterns are indeed everywhere. In this unit, students examine patterns and begin to analyze what relationships exist among the elements of patterns and how that information can be used to predict what might happen next.

One of the major ideas in this unit is that patterns are predictable. Patterns may have elements that alternate, repeat, increase, or decrease in a regular way. Once these regular relationships have been described, we can predict how the pattern will continue. As kindergarten students snap together a set of red and white cubes, they may begin to notice that they can alternate these colors to form a pattern: red-white-red-white-red-white. In time they will realize that this repeating pattern can go on indefinitely and that it provides a concrete model for fundamental aspects of our number system: the relationship of odd and even numbers, and the idea of infinity.

In this unit, students construct concrete models of patterns. In so doing, they sort, classify, count, and organize objects. As students make patterns, they begin to see the power in being able to make predictions about how a pattern continues. They also learn to identify when there is not enough information to make a prediction.

For most kindergartners, predictions will focus on what element comes next in a linear pattern. A few will be able to describe chunks or units of a pattern, but many will think only in terms of "what comes after what." That is, for the cube sequence yellow-red-yellow-red-yellow-red, students won't necessarily think about the pattern as being made from the unit yellow-red. Rather, they will think of the pattern as "red comes after yellow, yellow comes after red." They think about the action of building a sequence in order to understand the pattern. Once students are able to predict what comes *next* in a pattern sequence, they can use this information to predict what comes at some later point in a pattern sequence, even when they don't have all the intervening information.

When students are asked to describe concrete models of patterns, they must decide which feature or features are most important. Do they consider color? shape? number? or some combination of these? As kindergartners construct a simple repeating pattern with alternating yellow hexagons and green triangles, they may first read their pattern as yellow, green, yellow, green. On a second look, they may see that they could also read this pattern as hexagon, triangle, hexagon, triangle.

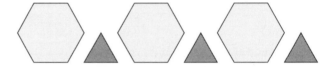

As students become more familiar with the structure of patterns, they begin to consider how this pattern of hexagons and triangles is both the same as and different from a pattern of alternating red trapezoids and orange squares.

Many young students will see these two patterns as unrelated because the most salient attributes of color and shape are different. Being able to identify the underlying similarity, that both are patterns of two alternating elements, is an important idea in understanding patterns.

Throughout this unit, students are learning to analyze the elements of a sequential pattern. A pattern sequence, like those that students make with interlocking cubes, is made up of repeating units. In the pattern red-white-red-white-red-white, we can label the elements of the sequence with letters, a-b-a-b-a-b. In this sequence the repeating unit is a-b, or red-white. Repetitions of this unit, one after the other, make a pattern.

a-b pattern

As units get larger and more complex, it is sometimes difficult to look at a pattern and figure out the basic unit. The pattern sequence a-b-b-a-b-b-a-b-b has a three element unit: a-b-b. However, the unit of the pattern sequence a-b-b-a-a-b-b-a-a-b-b-a is harder to see. Do you recognize it? This pattern shows three repetitions of the basic unit a-b-b-a.

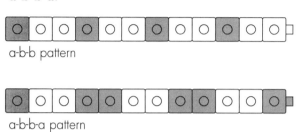

a-b-b pattern

a-b-b-a pattern

Being able to decompose a pattern into its repeating parts is an important mathematical idea. This will be a challenge to many young students, but as they have repeated opportunities to copy, create, extend, describe, compare, and discuss patterns, they will become more comfortable with finding the *unit* of repeating patterns.

Following their work with repeating patterns, students explore patterns that increase and decrease in a regular and predictable way. They build cube "staircases," focusing on the number of cubes in a step in order to determine how a pattern grows or shrinks—for example, by one cube each time, or by two.

Mathematics has been called "the science of patterns," for it is often used as a language to describe and predict numerical or geometrical regularities. Some of these patterns are very simple. One of the first numerical patterns that kindergartners notice is the "plus one" pattern of the counting sequence. They *hear* the pattern of the numbers repeating as they count, and when these numbers are arranged on a 100 chart, they also *see* patterns in the arrangement. These patterns are not accidental; they indicate important aspects of our number system and of mathematical relationships. While the focus of this unit is on helping students begin to see and construct patterns, these experiences become the foundation for later work in elementary school when students will explore patterns in number and geometry.

At the beginning of each investigation, the Mathematical Emphasis section tells you what is most important for students to learn about during the investigation. Many of these mathematical understandings and processes are difficult and complex. Students gradually learn more and more about each idea over many years of schooling. Individual students will begin and end the unit with different levels of knowledge and skill, but all will gain greater understanding of the regularity and predictability that characterize patterns.

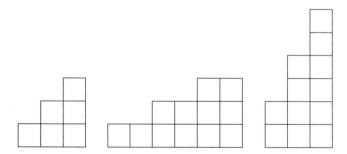

Staircase patterns

Throughout the *Investigations* curriculum, there are many opportunities for ongoing daily assessment as you observe, listen to, and interact with students at work. You can use almost any activity in this unit to assess your students' needs and strengths. Listed below are questions to help you focus your observations in each investigation. You may want to keep track of your observations for each student to help you plan your curriculum and monitor students' growth. Suggestions for documenting student growth can be found in the section About Assessment (p. I-8).

Investigation 1: Exploring Patterns

■ How observant are students? As they look carefully at their surroundings and at patterns and arrangements, what do they notice? What attributes (size, shape, color, function, position, proximity) do they observe and attend to?

■ What do students identify as patterns? Do they include designs? stencils? arrangements of pattern blocks? rhythms? How do they decide if something is a pattern or not? Do they recognize patterns in their environment?

■ How do students create patterns? Do they make random arrangements or arrangements that are predictable? What materials do they use? Are they comfortable making a variety of two-variable patterns (a-b, a-b-b, a-a-b, a-a-b-b) or do they create only one type of pattern (a-b)?

■ Can students copy a pattern? Can they add onto an existing pattern? How do they predict what comes next? Do they read the entire pattern, starting from the first element? from the last element? Do they "just know" what comes after what (yellow comes after red; red comes after yellow)?

Investigation 2: What Comes Next?

■ Do students recognize patterns? How do they decide if something is or is not a pattern? Can they explain their thinking? What attributes (color, shape, size, position, quantity) do they focus on?

■ How do students construct two- and three-variable patterns? What materials are they most comfortable using? Do they always make an a-b pattern? Do they consider red-yellow-red-yellow to be the same as or different from yellow-red-yellow-red? the same as or different from blue-green-blue-green? Are any students trying a-b-b patterns or a-a-b patterns?

■ How do students extend existing patterns? Can they describe how they know what comes next?

■ Can students "read" a pattern? What attributes do they attend to?

■ Are students able to record a pattern they have built? What recording strategies do they use?

■ Can students predict the color of any element (not necessarily the next) that is hidden from view? What strategies do they use for doing so?

■ Are students able to identify the unit of a pattern? How do they break a pattern down into individual units? How do they assemble a pattern given the units (or parts of units)? Can they figure out how many times the unit of the pattern repeats?

Investigation 3: Hopscotch Paths

- What type of patterns do students make with the large hopscotch squares? Can they extend their own pattern? another student's?

- Can students replicate their hopscotch patterns using color tiles? paper squares?

- How do students use physical movements to interpret patterns? Can they act out a pattern made from large hopscotch squares? from tiles? from paper squares? Can they act out a pattern using more than one motion or movement? Can they observe a physical pattern and represent it with materials?

- Are students able to record patterns they make with materials? Can they record patterns made with their bodies? How do they do so? What materials do they prefer to use?

- How do students predict what comes next in a pattern? Can they predict a hidden element of a pattern that is not next in the sequence? How?

Investigation 4: Pattern Borders

- What type of patterns do students construct in border frames? Are they beginning to use more complex patterns in addition to a-b patterns? Are students able to turn each corner of the border and continue the pattern in a new direction? Can students predict what color tile comes next in their pattern?

- Are students able to use the information they have about a pattern to make predictions? If you point to an empty square anywhere on their border, can they predict what color that square will be? How do they predict this color?

- Are students able to make a variety of patterns using the same two colors (a-b-b, a-a-b, a-a-b-b, a-a-a-b), or are they able to make only an a-b pattern? If students are having difficulty making more complex patterns, are they able to extend a pattern such as a-b-b that you start for them?

- Can students double-check their pattern to see if it is correct? If they have made an error in their pattern, can they find what is wrong and fix it?

- Are students able to make any generalizations about the patterns they are making? For example, can students identify patterns as being the same type when they have the same structure but different variables (such as red-yellow-red-yellow and blue-green-blue-green)?

- Are students able to copy the staircase patterns accurately? Can they extend the pattern going up and going down? Can they predict what step will come next in the staircase? How do student describe the staircases? with words? with numbers? Do they recognize the pattern of the staircase? If so, how do they describe that?

- Can students transfer their work to a grid to record it? How do they do this? Do they remove one square at a time and color in that square? Do they place the recording sheet next to their cube towers, using the concrete pattern as a reference? Do students seem to internalize their pattern and just color it in without referring to the original model?

In the *Investigations* curriculum, mathematical vocabulary is introduced naturally during the activities. We don't ask students to learn definitions of new terms; rather, they come to understand such words as *triangle*, *add*, *compare*, *data*, and *graph* by hearing them used frequently in discussion as they investigate new concepts. This approach is compatible with current theories of second-language acquisition, which emphasize the use of new vocabulary in meaningful contexts while students are actively involved with objects, pictures, and physical movement.

Listed below are some key words used in this unit that will not be new to most English speakers at this age level, but may be unfamiliar to students with limited English proficiency. You will want to spend additional time working on these words with your students who are learning English. If your students are working with a second-language teacher, you might enlist your colleague's aid in familiarizing students with these words, before and during this unit. In the classroom, look for opportunities for students to hear and use these words. Activities you can use to present the words are given in the appendix, Vocabulary Support for Second-Language Learners (p. 102).

missing, same, different Students first sharpen their observation skills with a game that involves identifying an item missing from a tray. Then, as they begin their exploration of pattern, they observe patterns of colored cubes to see how they are the same and how they are different.

color names Throughout the unit, students work with pattern sequences based on color, so they need to be familiar with the names for the colors of the interlocking cubes, the pattern blocks, and other pattern-making materials you are using.

next, add As students continue to work with linear repeating patterns, they begin to identify what comes next, and add that element to the pattern.

squares Students need to recognize the word *squares* in their work with patterns of color tiles and paper squares of different sizes.

corner, edge For their work on pattern borders in Investigation 4, students discover what happens to a pattern as it goes around all the edges and turns the corners of a square.

up, down In the activity Staircase Patterns, students work with sequences of cubes that go up and down like a staircase.

Multicultural Extensions for All Students

Whenever possible, encourage students to share words, objects, customs, or any aspects of daily life from their own cultures and backgrounds that are relevant to the activities in this unit. For example, help students explore patterns that are present or popular in different cultures, perhaps in cloth, clothing, or traditional songs and dances.

Three children's books that explore patterns in different cultures are *Ten Little Rabbits* by Virginia Grossman (Native American blankets), *Kente Colors* by Debbi Chocolate (the symbolic colors and patterns of the Ashanti and Ewe people of Ghana), and *Jyoti's Journey* by Helen Ganly (full of patterns from India).

Investigations

Exploring Patterns

Focus Time

Watching and Looking (p. 4)

To introduce this unit, students practice being observant as they play a "do what I do" game, share a book about seeing their surroundings, and take an observation walk. Looking at things carefully and noticing their attributes is an important part of recognizing and constructing patterns.

Cubes: What Do You Notice? (p. 8)

Students next observe trains of colored interlocking cubes, some arranged in patterns and some not. They compare what they notice about each arrangement and predict what color might come next.

Choice Time

Making Patterns (p. 14)

Students construct linear patterns with a variety of materials such as pattern blocks, color tiles, interlocking cubes, and teddy bear counters.

What's Missing? (p. 16)

In this game, students closely observe a set of objects. When one or two objects are removed, students try to identify what's missing.

Mathematical Emphasis

- Observing and describing attributes
- Recognizing and describing a pattern
- Creating and extending patterns using a variety of materials
- Predicting what comes next in a pattern

Teacher Support

Teacher Notes

Patterns and Children's Literature (p. 18)

What Does "Pattern" Mean? (p. 19)

Choosing Materials for Pattern Activities (p. 20)

Dialogue Boxes

Two Arrangements of Color (p. 21)

Sharing Our Cube Trains (p. 22)

What Might Come Next? (p. 23)

What to Plan Ahead of Time

Focus Time Materials

Watching and Looking

- *I Am Eyes • Ni Macho* by Leila Ward (New York: Scholastic, 1978), or similar book about observing your environment (optional)

Cubes: What Do You Notice?

- Interlocking cubes: 1 tub per 4–6 students
- Two 8-cube trains for demonstration, one train made with eight different colors, the other train with just two colors in an alternating a-b pattern

Note: If your students have not had previous experience with interlocking cubes, plan some time for free exploration of this material before introducing the activity Cubes: What Do You Notice? For information on free exploration see the **Teacher Note**, Materials as Tools for Learning (p. 85).

Choice Time Materials

Making Patterns

- A variety of manipulative materials, such as interlocking cubes, pattern blocks, color tiles, teddy bear counters, keys, and buttons. See the **Teacher Note,** Choosing Materials for Pattern Activities, p. 20.
- Pattern Paths (p. 105): 1 per student, cut and glued as indicated on blackline master (optional)

What's Missing?

- Small trays or sheets of construction paper, about 12 by 18 inches
- Pieces of covering material as large or larger than the trays (towels, dark scarves, fabric remnants, or large paper)
- Masking tape or wide-tip markers (for marking tray divisions)
- Small paper bags
- Collections of 9–12 different small objects, such as pattern blocks, cubes, counters, buttons, shells, rocks, keys, crayons, and other classroom objects

Family Connection

- Family letter (p. 104): 1 per family

Watching and Looking

What Happens

Students play a pattern game called *Can You Do What I Do?* In this game they try to copy the motions of the teacher's hands and body. The group reads a book that focuses on being observant and aware of one's surroundings. The class then goes on an observation walk around the school or neighborhood. These initial activities prepare students for looking for and creating patterns. Their work focuses on:

■ following a routine or pattern of body motions

■ creating a routine or pattern involving body motions

■ predicting what comes next in a routine or pattern

■ observing and describing what you see

Note: Looking carefully at how things are related and identifying attributes are important ideas in recognizing and constructing patterns. The first three activities of this investigation involve students in gathering information through observation. While the activities do have connections to pattern making, the main emphasis is on being observant.

Materials and Preparation

■ Obtain a copy of *I Am Eyes • Ni Macho* by Leila Ward, in big book form if possible, or a similar book about being observant and noticing your surroundings, such as *Brown Bear, Brown Bear, What Do You See?* by Bill Martin, Jr. (Holt, 1992) or *I Went Walking* by Sue Williams (Harcourt Brace Jovanovich, 1990).

Can You Do What I Do?

Position yourself where the whole group can easily see you. Students will need ample space to be able to move without disturbing each other. Some teachers find that sitting in a circle works best.

Begin a repeating, two-part gesture that involves both hands, such as *tap your shoulders, tap your head.* Your motions should have a steady beat to them. Ask students to join in and do the actions with you.

Can you do what I do? Can you do it, too?

Continue this routine until all students are moving with you. If you notice that the gesture you selected (such as snapping fingers) is too difficult for your students, quickly change to another motion. Following along may be difficult for some students. Don't be too quick to move on. Some teachers find it helpful to put words to the routine, for example, "Touch your shoulders. Touch your head. Touch your shoulders. Touch your head." However, if you can avoid doing this, you allow students to come up with their own words to describe the routine.

During this first session try to do three or four different types of routines. Use only a two-motion repetition at this point. You might include clapping gestures, standing and sitting movements, or tapping various parts of your body (knees, toes, head, shoulders, and so forth).

As you finish each routine, ask students to consider the following:

How do you know what to do? . . . How could you predict what might come next? . . . How would you describe to someone else what we did?

As you continue to play this game later in this unit, when students seem to understand what a *pattern* is, you can also ask them to consider this:

Did we create a pattern? How can you tell?

Once this activity has been introduced, students can play it in pairs or small groups. Some teachers use the *Can You Do What I Do?* game as a classroom routine during transition times or for getting students' attention. It is important to spend some time periodically asking students to focus on the repetition of each routine and discussing how they know what comes next.

A Book About Looking

When you were playing this game, I noticed that you were watching me very carefully to see what I was going to do next. You were being very *observant.* Can you think of some reasons why people need to be observant, or to notice things very carefully?

Collect ideas from the students. Encourage them to clarify or extend their ideas about why it is important to be observant. Explain that you are going to read them a book about being observant and looking very carefully at things in the world.

The suggested book for this activity, *I Am Eyes • Ni Macho,* is a simple story that introduces the idea of being observant and being aware of the environment. The alliterated text begins with a young African boy waking to a bright, sunny day. He declares, "Ni macho!" which means to be awake, or literally, "I am eyes." Each page begins "I see . . ." and is followed by an alliterative phrase such as "sunflowers and skies," or "flowers and flamingos."

Read this book aloud or use another story to create a similar context. See the **Teacher Note,** Patterns and Children's Literature (p. 18), for more examples of weaving together literature and mathematics.

After reading the entire book, turn back and ask students to look carefully at some of the pages.

What sorts of things did the boy notice as he was walking? What are some things that you notice about this page? about this book?

Students are likely to comment on the objects or animals in the pictures. Some may be interested in the border design that goes around each page. Others may comment on the repeating pattern in the words. You may want to choose one observation and look to see if the same element appears on other pages throughout the book.

Observation Walk

After the group has shared observations about the book, explain that just like the boy in the story, the class is going for a walk. If possible, take a walking trip around your school or playground. Otherwise, plan an observation walk around your classroom, maybe breaking into smaller groups.

As we take our walk, I'd like you to use your eyes and look very carefully at what is around you. Try to notice as much as you can. There may be things you have never noticed before. Try to remember a few things to share with the group when we return.

As you walk, focus students' attention on specific things, especially interesting arrangements of objects or patterns that appear in the surroundings. The main purpose is to get students looking in new ways at their environment.

During the walk, encourage students to point out what they are noticing. When you return, make quick sketches to record their observations on chart paper or have students themselves draw a picture of things they noticed on their walk.

Cubes: What Do You Notice?

What Happens

As a whole group, students look at two different trains of interlocking cubes, one with eight colors arranged randomly and one with two colors in a repeating a-b pattern. They compare what they notice about the two arrangements and try to predict what color comes next. Students then create their own arrangement of cubes and, as a whole class, sort these arrangements into two groups. Their work focuses on:

- noticing and describing attributes of an arrangement (size, color, shape, quantity)
- describing similarities and differences between two arrangements
- predicting what comes next in a color sequence
- making arrangements of color with cubes
- distinguishing between a random arrangement and a predictable arrangement or pattern

Materials and Preparation

- For demonstration, make two trains of eight interlocking cubes. In one train, use eight different colors randomly arranged; for example, red-white-black-blue-orange-yellow-green-brown. In the other train, use two colors arranged in an alternating (a-b) pattern; for example: red-white-red-white-red-white-red-white.
- Make available a tub of interlocking cubes for each 4–6 students.

Two Arrangements of Cubes

Note: Be sure that students have had the opportunity to freely explore the interlocking cubes before you proceed with this activity. See the **Teacher Note**, Materials as Tools for Learning (p. 85).

I have made two different cube trains. Today we are going to look very closely at these trains to see how the cubes go together and how they are arranged.

Hold up the train of cubes with eight colors arranged in random order.

What do you notice about this?

Student responses will vary. Some will be quick to talk about color. Others will notice the way the cubes interlock. Still others will count the number of cubes used. Gather many different responses, and then show students the second train of cubes with two colors arranged in an alternating a-b pattern.

Here's a different train of cubes. What can you tell me about this one?

We suggest that you *not* introduce the word *pattern* to students right away, but instead offer them opportunities to observe different arrangements of cubes and describe what is the same and what is different about those arrangements. For many kindergarten students, *pattern* may not be a familiar vocabulary word, even though they may recognize and accurately describe the attributes of a pattern.

So, rather than pointing out that the cubes in the second train are arranged in a pattern, listen to the way students themselves describe this arrangement. After gathering many responses, ask students to compare the two sets of cubes.

We have been looking carefully at these two cube trains. What do you notice is the *same* about them? What is *different*?

The **Dialogue Box**, Two Arrangements of Color (p. 21), offers examples of observations made by some kindergarten students.

Observing the Students

Listening to the responses students give is important. The following questions can help you gauge what students are telling you through their observations.

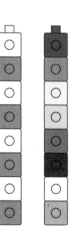

- How do students describe the arrangements? Are responses personal? ("I like it.") Are they imaginative? ("It looks like a candy cane.") Are they descriptive? ("It has lots of colors," or "It has eight squares.") Does anyone use the word *pattern* to describe either arrangement?
- How accurate are their statements? Are descriptive words (size, color, shape, quantity) used correctly?
- Are students making connections to statements made by other students? to prior experiences?
- How are students responding to the discussion? Do they need to come up and show rather than tell you what they see? Are their ideas original?

Making Cube Trains

Explain that students will be making their own trains of interlocking cubes, similar to the ones they have just seen.

Every train you make should have eight cubes, just like my cube trains. Your trains can look like one of these trains that we have been talking about, or they can look different. At the end of our work time, you will choose one of your cube trains to share with the group.

Send small groups of students to work at the tubs of cubes set out around the classroom. See the **Teacher Notes,** What Does "Pattern" Mean? (p. 19) and Choosing Materials for Pattern Activities (p. 20), for background information on introducing kindergartners to patterns.

Observing the Students

As students are working, circulate around the room and ask them to describe their arrangements of cubes. You might carry with you the two cube trains that you displayed during meeting so that students can think about whether their trains are similar or different.

- Are students able to make a train with exactly eight cubes? If not, ask them to separate the cubes in their train and count them. Sometimes it is easier for students to count cubes individually and then snap them together rather than to count them in a connected train. If students are still having difficulty, put out eight cubes and ask the students to match their cubes, one-to-one, to that set.
- Do students make random arrangements of color or arrangements that are predictable?
- What are students saying about the similarities and differences among arrangements of color?

At the end of this work session (about 20 minutes), ask students to choose one of their cube trains to share with the class. They should break apart all other cube trains and return the cubes to the tubs.

Sharing Cube Trains

Note: Discussions in which students share their work and their strategies are an important part of the *Investigations* curriculum. For more information, see the **Teacher Note,** Encouraging Students to Think, Reason, and Share Ideas (p. 86).

Gather students together in a whole group where they can see each other's cube trains.

Take a careful look at all the different trains of cubes you have made. What do you notice about the different arrangements?

After a few students have shared observations, suggest to them a way of grouping the trains by using the two original trains you showed.

Let's organize your trains into two piles. Did anyone make a train similar to this one, with two colors? If you did, please put your train here in this pile. How about this other arrangement? Did anyone make a train with lots of colors? If so, put your train in this other pile.

As students place their trains into the two piles, organize them so that everyone can see them. If students are unsure where their train belongs, make suggestions so that this part of the activity moves along fairly quickly. If there are arrangements that aren't clearly one type or the other, place them in a third pile.

Look very carefully at each group of trains. Which arrangements make it easy to tell what color comes next?

As you gather responses, ask students to explain their reasoning. Listen for evidence that might suggest they have identified the predictable nature of a pattern arrangement. See the **Dialogue Box**, Sharing Our Cube Trains (p. 22), for an example of how this discussion unfolded in one kindergarten classroom.

What Might Come Next?

Choose one of the trains from the group with colors randomly arranged and ask students what they think might come next if you added another cube to the end of the train. Some teachers have found that making cubes available during this activity encourages more students to be involved. It also enables the teacher to assess individual students in a group discussion.

Student responses will vary because in the random arrangements, there is no way to predict what might come next. The **Dialogue Box,** What Might Come Next? (p. 23), shows how the discussion went in one class. Continue the discussion long enough to allow a range of ideas to emerge. Then hold up a train from the group with two alternating colors (a-b pattern).

If I were to add another cube to the end of this train, what color might go here? Why do you think that?

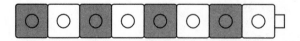

Students might observe that it is easier to tell what comes next in this arrangement. You may hear comments like, "Oh, it has to be a red," or "It can only be red or white."

What makes you so sure? Why is it easy to tell what comes next in this arrangement, but not so easy in the first one we looked at?

Help students recognize the predictable nature of the two-color alternating arrangement. If at this point no one has called this arrangement a *pattern*, introduce the word now. Point out which trains have repeating color patterns and which trains have no color pattern. The **Dialogue Box**, Two Arrangements of Color (p. 21), demonstrates how one teacher first talked about the word *pattern* with students.

Focus Time Follow-Up

Family Connection Send home the signed family letter or the *Investigations* at Home booklet to introduce the work you will be doing in this pattern unit.

Two Choices If Choice Time is not already a standard part of your kindergarten program, refer to the **Teacher Note**, About Choice Time (p. 82), for more information.

The following independent activities support students' work with observing closely and with making simple linear patterns. On the first day of Choice Time, introduce Making Patterns (p. 14); then on the following day, introduce What's Missing? (p. 16) as another choice.

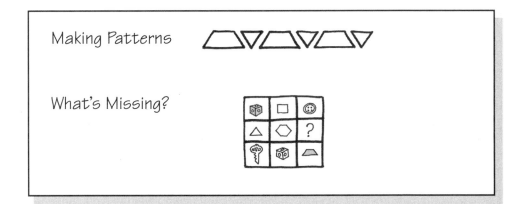

Making Patterns

What Happens

Students use a variety of materials to construct their own linear patterns. Their work focuses on:

- constructing patterns
- thinking about what might come next in a pattern
- extending patterns
- discriminating between a pattern and an arrangement or design

Materials and Preparation

- Set up stations with two or three different materials, such as interlocking cubes, pattern blocks, color tiles, teddy bear counters, keys, and buttons. See the **Teacher Note**, Choosing Materials for Pattern Activities, p. 20.
- Duplicate Pattern Paths (p. 105), 1 per student (as an option). Assemble with scissors and glue, making a row of twelve squares for each path. If possible, copy on card stock and/or laminate. Plan to save these for further use in Investigation 2.

Before You Begin

When students are asked to use materials such as pattern blocks, color tiles, or teddy bear counters in a structured way, it is important that they have previously had the chance to play with and freely explore the materials. If you have used the *Investigations* unit *Mathematical Thinking in Kindergarten* with your class, students will have already had this free exploration time. Otherwise, you will need to offer similar experiences before asking students to make patterns with the materials.

Activity

Choose two or three different materials for students to work with and place them on different tables around the classroom. Every few days, vary the selection so that students have the opportunity to create patterns with a variety of materials. Select one type to introduce the activity.

Today we are going to use the colored cubes and also some new materials to make patterns and arrangements.

Suppose I put out a yellow hexagon, then a green triangle, then a yellow hexagon, then a green triangle. What comes next in my pattern? How do you know?

Invite a few students to add a block on to your pattern. Ask students to explain why they think the block they chose comes next.

Show students the Pattern Paths you have assembled.

When you are building a pattern, you might want to use this path as a mat to build on. You can place one object in each square of your path. Once you have filled a Pattern Path, you might ask someone at your table to guess what comes next.

Ask a volunteer to demonstrate how to build a pattern on a Pattern Path. These mats are a helpful organizational tool for some students, but others may find it easier to "see" their pattern by placing the objects closer together. They should understand that use of the Pattern Paths is optional.

Observing the Students

Making patterns may be a challenging task for some students. As you observe students, ask them to tell you about their work. Consider the following questions.

- Are students making patterns or are they arranging items randomly? What type of patterns are they making? Is there variety in their work?
- If students are unable to construct their own pattern, are they able to add on and continue a pattern that you start for them?
- How do students "read" their pattern? How do they describe the elements in the pattern? by size? color? shape? quantity? position?
- Can students predict what comes next in a pattern? Do they read the pattern from the beginning each time, or can they tell from the last few elements what comes next?

Variation

Make available art supplies, such as rubber stamps, stickers, markers, or paints, for making patterns on unlined paper.

What's Missing?

What Happens

Students play What's Missing? in small groups. They closely observe a set of objects on a tray before it is covered with a cloth. After a student "magician" makes one or two objects disappear into a bag, the other students try to identify what's missing. Students' work focuses on:

- being observant
- using information to figure out what is missing

Materials and Preparation

- Prepare gameboards from small trays or sheets of construction paper, about 12 by 18 inches. Mark each gameboard in 9 sections (as for tic-tac-toe). Use masking tape or wide-tip markers for the dividing lines.

- Find a piece of material to completely conceal each gameboard (consider towels, dark scarves, fabric remnants, or large paper).

- Prepare collections of 9–12 different small objects. For example, you might use classroom objects (crayons, markers, buttons, pattern blocks, counters) or things from nature (acorns, shells, rocks, sticks, seeds).

- Set up three or four stations, each with one gameboard (prepared as described above), a covering sheet, a small paper bag, and one of the collections of 9–12 objects.

Activity

Introduce this game to the whole class or to a small group. Show students each of the collections available at the stations, but choose one set for demonstrating the game. Place nine objects in the sections on the gameboard and display it where everyone can see.

Look very carefully at the items I have on my tray. In a minute I am going to cover them up. Then I'll take one or two away and put them in this paper bag. Your job will be to figure out what I took away.

After all students have had an opportunity to look closely at the items (about 1 minute), cover the gameboard with a piece of fabric or paper. Ask students to hide their eyes while you remove one or two objects from the tray. Hide these objects in the paper bag. Uncover the gameboard and ask the students which items are missing and how they know.

Explain to students that during Choice Time, they can play this game with two or three other people. One person will be the "magician" who makes things disappear from the tray, and the other players guess what's missing.

Students in a group take turns being the magician. Encourage them to use the same nine objects for at least a few rounds of the game, so that they become very familiar with what is in each set. Some magicians might want to rearrange the objects on the tray (after removing one or two) as a way of challenging their teammates.

Observing the Students

Consider the following while you observe students playing What's Missing?

- What strategies do students use to figure out the missing object(s)?
- Do students attend to attributes, such as the size, shape, color, or function of objects, to help them determine what is missing?
- Do they use position or proximity?
- Can students explain their reasoning, logic, or strategies for determining what is missing?

Variation

Once students are familiar with the collections of objects you have prepared, replace some of the objects with new ones. Students can also be given the task of assembling new sets of objects.

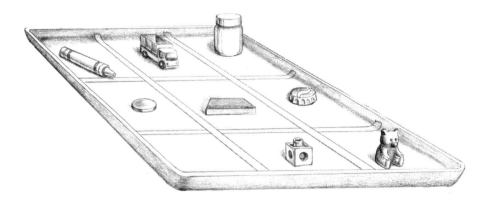

The study of patterns easily branches out beyond the world of mathematics. It is a part of art, architecture, construction, science, economics, medicine, and more. Language and literature are filled with examples of patterns.

Numerous children's books employ repetitive language and predictable texts, giving you opportunities to weave language arts and mathematics together. Teachers commonly report students saying "That's a pattern!" as they listen to a story being read aloud. Sometimes the children are identifying an element in the story or an illustration that is indeed a pattern. Such is the case when a student notices the footprints Peter leaves in the snow in the classic *The Snowy Day* by Ezra Jack Keats (Viking, 1963): "Look, Peter walks with his toes pointing in and then with them pointing out." Some will even notice mathematical patterns, such as the numeric sequence of eating "one more thing each time" in *The Very Hungry Caterpillar* by Eric Carle (Philomel Books, 1981).

Sometimes students believe a pattern exists even though one does not. This was the case in one class when students noticed that the first page of the book *I Am Eyes • Ni Macho* by Leila Ward showed one boy. They predicted that two boys would be on the second page. As the page was turned, students found their prediction was not confirmed. They began to make more predictions.

"Maybe it will always show just one boy."

"No, I think it will go 1, 1, 2."

In each case, the students found that their predictions were unfounded. Even after making a list of the number of people on each page of the book, no pattern was found. This was disturbing to these 5- and 6-year-olds. They had felt an urge to make sense of the arrangement in these illustrations and they weren't satisfied with the outcome.

The teacher reported that days after this, some students were still looking for patterns in the illustrations. One day a student announced: "There *is* a pattern, it's just not with the people. It goes [showing the book] one page is black and one page is color, one page is black and one page is color." With some satisfaction, the students in this kindergarten class felt that a pattern had at last been discovered.

Patterns in literature can often be found in illustrations or in the language of the text. Many books for young readers are "predictable," with words or phrases that regularly repeat. Many kindergartners recognize this repetition and refer to these as patterns. The repetition guides the emerging reader in making predictions about what comes next. In the story *Brown Bear, Brown Bear, What Do You See?* by Bill Martin, Jr. (Holt, 1992), the familiar question of the title repeats, followed by "I see a [color word + animal] looking at me." A pattern is quickly established and followed throughout the book.

Highlight times when you notice patterns in stories being read aloud and encourage your students to do the same. These literary investigations are a wonderful way to connect mathematics to other aspects of the curriculum.

What Does "Pattern" Mean?

As students start to build patterns with manipulatives, they often label any configuration they have generated or any design they have built as "a pattern." How do students begin to develop meaning for the word *pattern*? How do they learn to distinguish between any design and an arrangement that is regular and predictable in some way?

The term *pattern* has multiple meanings. One teacher heard a girl refer to her pattern block construction of a dog as a "pattern."

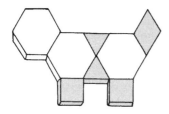

Later that day, the same girl asked a classmate in the art center to "please pass the pattern of the bear."

At this moment, the teacher realized that some students in her class were referring to stencils as *patterns*. Later she asked the girl what she thought a pattern might be. "Oh, you know, it's like when my mom is making a dress and she has a pattern." Hearing this, the teacher understood why the girl called the stencil a pattern. But thinking back to the first situation, the teacher felt confused about the girl's thinking. Was she thinking of the dog she created with the pattern blocks as a stencil-type pattern? Or did the name of the material, *pattern* blocks, suggest to the girl that anything she built with those blocks was a pattern? This illustrates the complexity of language that young students are acquiring and deciphering, and for which they are negotiating meaning. The language of mathematics adds yet another layer of complexity for students to explore and make their own.

Sometimes teachers elect to use the terms *arrangement* or *design* to label constructions that do not have a discernible repetition or regularity, such as the girl's pattern block dog. They reserve the term *pattern* to label or model work that clearly shows repetition or regularity. By using language carefully, teachers can model ways of describing what students are doing without directly correcting them. To learn about students' thinking, avoid passing judgment on whether or not something is a pattern; instead, ask students to describe their work to you and to their classmates, explaining why they think it is (or is not) a pattern.

The goal of this unit is to expand and refine students' notions of what constitutes a pattern. Listening to and observing students in order to make sense of their understanding is at the heart of this teaching.

A wide range of manipulative materials are produced specifically for students to use as they explore mathematical concepts. Other materials that can be collected around the classroom and at home—such as buttons, shells, keys, and rocks—are also useful for mathematical explorations. As you examine the materials in your classroom, consider which collections best lend themselves to an investigation of patterns.

Students generate patterns by isolating one or more attributes of a given material. For kindergarten students, color, size, and shape are often the most prominent features; they may sometimes use quantity, purpose, and position as classification categories. Snapping together a set of interlocking cubes by alternating their colors is easier for most students than laying out a sequential pattern of buttons based on the number of holes or button size. In selecting materials, look for distinguishing features that students can use to begin thinking about patterns. Some materials are more supportive of their learning than others.

Even when the focus is on patterns, expect that students may continue to build or sort with some of these materials as well. Children are naturally inclined to build and to sort items into like groups. With enough time for exploratory play, even young students will eventually be able to focus their work in the way you ask them.

Occasionally teachers will ask if and when they should limit the number of items used to create a pattern. Children 5 and 6 years old enjoy stringing items together to make long trains. It is not uncommon for students to challenge themselves to see how far across the classroom they can run a train of interlocking cubes, or how large an area they can cover using yellow hexagons from the pattern block set. These challenges help them gain some understanding of measurement, estimation, and quantity.

During the investigations on pattern, you can encourage students to make long trains at first so they encounter the idea that a pattern could continue indefinitely. However, as the activities in this investigation begin to focus on similarities and differences among patterns, very long trains are cumbersome and use a lot of materials. Providing defined spaces or paths for building patterns will help some students limit and focus their work. Pattern Paths offer a framework for linear patterns. Similarly, a sheet of paper used as a mat provides a clear area within which other kinds of patterns, such as border or quilt designs, can be created.

In kindergarten, we suggest that students work mostly with a-b patterns and other patterns made with two elements, such as a-b-b, a-a-b or a-a-a-b-b-b. However, introducing a third element may be appropriate for some students. Giving students constraints to work with often brings out important ideas that might otherwise go unnoticed. The constraints can be as simple as working with a specific number of objects, constructing patterns using just two elements, or creating a design that fits within a certain space.

For example, in the activity 12 Chips (p. 76), students make different patterns using a total of 12 red and yellow chips. After building each pattern, they record how many red and how many yellow chips they used. Students may begin to describe why there are the same number of reds and yellows in a red-yellow-red-yellow-red-yellow pattern, and why there are more reds than yellows in a red-red-yellow-red-red-yellow pattern.

Working on activities with constraints should always be balanced with the chance to use materials in ways that students choose. In both situations, students discover interesting and important things about the materials and about the mathematics involved.

Two Arrangements of Color

During Focus Time (Cubes: What Do You Notice? p. 8), students discuss two different cube trains, one with random colors, the other with colors arranged in a repeating a-b pattern. At this point in the year, students have not been formally introduced to the word *pattern*. The teacher is interested in seeing how students will describe the two trains and if any will use the word *pattern* to describe the arrangement of colors that repeats. To learn more about what they are thinking, the teacher encourages students to explain their observations.

What do you see in this train *[holds up the one with random colors]*?

Thomas: It looks like a rainbow.

Renata: No. Rainbows don't have black and white.

Maddy: I see eight squares.

Tarik: They make a train. See, each block is like a car. The red is the engine. The brown is the caboose.

Jacob: I see that too, but I think there are nine cars.

Raise your hand if you were thinking about the number of cubes I'm holding . . . I see that some of you were thinking about the number. Let's count them together.

All students: 1, 2, 3, 4, 5, 6, 7, 8.

Maddy: Yup, there's eight.

Tarik: Eight cars in the train!

Now look at this other train *[with alternating colors]*.

Tiana: It has only two colors.

Alexa: I see red and white, red and white.

Maddy: It still has eight.

Tarik: It's still a train with eight cars.

Oscar: It's like my shirt. *[He is wearing a checkerboard flannel shirt.]*

Miyuki: It goes red, white, red, white, red, white, red, white. *[She points to the cubes the teacher is holding, moving her finger from left to right as she says each color and goes on to the next cube in the train.]*

How would you compare these two trains? What's the same, and what's different?

Ravi: That one has a lot of colors. That one has only two colors.

Charlotte: I think this *[points to the red and white train]* is a pattern.

What do you mean it is a pattern?

Charlotte: A pattern keeps going over and over. See red, white, red, white. The other one isn't a pattern because it's just colors.

Felipe: One is pretty and one is striped.

Can you tell us more about what you mean?

Felipe: I think this one is pretty *[picks up the train with multiple colors]*. It has lots of colors. It has my favorite color green. This one *[reaches for the second train]* has stripes. It goes red, white, red, white. . .

Tess: One is a rainbow and one is a candy cane.

Charlotte has used the word *pattern* to describe this type of arrangement. We are going to be learning a lot about patterns in the next few weeks. This *[holds up the red and white train]* is a pattern, and this *[holds up the train with random colors]* is not a pattern. Has anyone else ever heard the word *pattern*?

Pattern is not a familiar word to most 5- and 6-year-olds, or if it is, they have a very limited understanding of its meaning. Now that the term has come up in discussion, the teacher will model its use as the class does the rest of the activities in this unit. In future discussions she will encourage students to think about what makes a pattern a pattern.

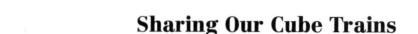

Sharing Our Cube Trains

During Focus Time, after working for about 20 minutes to make their own cube trains, students each bring one cube train to the meeting area.

Carlo: I made mine just like the rainbow one, but I used purple.

Justine: I made mine with only two colors. See, black and white like a zebra.

Ayesha: I used two colors, blue and yellow.

I notice that several of you used just two colors. Let's put those together in a pile so we can look at them more closely. Let's also make a pile of those with more than two colors. Will we need any other piles? *[The class agrees that two piles will work.]* **Is there anything we can say about the two piles? What do you notice?**

Kadim: More kids used rainbow colors.

Gabriela: Lots of kids made red and white ones like candy canes.

Let's look at just one pile at a time. What do you notice about these *[with random colors]*?

Felipe: I like the rainbow ones. They're pretty.

Luke: They are all different colors.

Now let's look at these. *[Points to the other pile.]* **What do you notice?**

Xing-Qi: They look like stripes.

Thomas: They look like a funny checkerboard.

Gabriela: Some of us made the same. There are a lot of candy canes.

Tess: These are all patterns. Well, this one isn't *[picks up a train that is red, white, red, white, red, white, white, red]*.

Can you make it a pattern?

Tess: *[Changes the order of the last two cubes]* Now it is. See, red, white, red, white. . . .

Tess just changed the order of some of the cubes. What do other people think? Does that make sense to you?

Kylie: I think it looks better now.

Brendan: That was mine. I meant to do that. I meant it to go red, white, red, white all the way down. I just missed.

What is it about this train that makes everyone so sure the last two blocks should be turned around?

Gabriela: It makes the candy cane.

Henry: You just know it has to be that way. Red, white, red, white, over and over and over.

It sure does go over and over! Which trains here go over and over and over? *[The class agrees that all the trains in the second pile go "over and over."]* **Which arrangements make it easy to tell what comes next?**

Gabriela: These do *[the second pile]*. They're stripes. *[Picks up a red and white train.]* You can see that white comes after red every time.

Justine: You can always tell what color comes next by looking for that color. Then it has to be the other one.

What about these? *[Points to the other pile.]*

Xing-Qi: No, they are just different colors.

Tess: They could be any color.

Gabriela: They don't make stripes.

Justine: You can't tell what comes next after black here *[picks up one random arrangement ending in black]* because there is no other black. How would you know?

Everyone agrees that the random colors don't help you know what comes next, and that the alternating colors do. In this discussion, the teacher helped students focus on the attributes of a pattern by offering them a structure for sorting the trains and asking them to compare and contrast them. While most students did not leave the discussion with a clear understanding of pattern, the foundation for future work and for future conversations was established.

What Might Come Next?

As the students again look at two arrangements of colored cubes—one random, one in a repeating pattern—they are asked to hold up a cube showing what color they think comes next in each arrangement. Students begin with a random arrangement (red-white-black-blue-orange-yellow-green-brown) and try to predict what color would come after the brown cube.

Ida: I think pink. I like pink and it's not there.

Shanique: No, red. See, it will start over.

Henry: *[Looking at his cubes on the floor in front of him]* It could be any color. It doesn't need to start over. That's boring.

Alexa: I don't know what goes here [the brown end], but I think white goes here *[she snaps her white cube to the left of the red cube].*

Why do you think white goes here?

Alexa: Because now it will go white, red, white just like the other one.

It is interesting to think of what might go here *[points to the cube Alexa has added to the arrangement],* **but I don't want to forget what the original train looks like, so I'm going to take this off and give you the white cube back.** *[Redirects the focus to the other end of the train]* **Any other ideas of what goes here after the brown cube? I see that Jacob is showing a yellow cube.**

Jacob: I like yellow.

After everyone had a chance to share their ideas, the teacher directed attention to the alternating red-white arrangement.

What might go here? Any ideas? Hold up one of the cubes from your cup to show what color you think comes next.

Carlo: I think it is going to be a red?

Tiana: Me too.

Miyuki: Not a red, a white.

Kadim: Yeah, white.

Thomas: But I want it to be blue like the flag. See—red, white, and blue! *[Points to the flag hanging near the meeting area.]*

Carlo: But it has to be red.

Miyuki: Or white!

Many of you are holding up a red or white cube. What makes you so sure?

Kadim: Because it has to be red, white, red, white. . . .

Carlo: Yup, it's got to be!

When we looked at the first arrangement, we had a lot of different ideas. For this arrangement, we seem to have just three ideas. Why is this? Who has an idea?

Ravi: The first one could be any colors. The second one can only be red or white. It's easy to tell what comes next because it has to be red or white. That's all.

Tiana: Yeah, red or white. The first one could be anything, even a color we don't know.

Shanique: But we could start the rainbow one over. *[Points to the first train.]* You could make this a pattern by starting over with red.

Brendan: But it's not a pattern now. Just the candy cane is a pattern. You can always tell what will come next.

Some of you are pretty sure that in a pattern you can always tell what comes next. This is important to think about. We could make the first group of cubes, the rainbow one, a pattern like Shanique said, but it would have to start over.

What Comes Next?

Focus Time

Patterns on the Pocket Chart (p. 26)

As students build their understanding of what a pattern is, they make and extend color patterns on the pocket chart and with color tiles and predict what comes next in a pattern when only part of the pattern is visible. Through this activity, students begin to look at the *unit* of a pattern, or the part of a pattern that repeats.

Choice Time

What Comes Next? (p. 32)

In this game, a player makes a pattern and hides the last six elements of it. The other player then builds the same pattern, copying as much as is showing, and predicts what's hidden.

Pattern Block Snakes (p. 34)

Students make linear patterns with pattern blocks and record their patterns by gluing paper pattern blocks onto paper strips.

Add On (p. 36)

In this game, players roll a number cube (with numbers 0–2) to determine how many cubes to add to a two-color pattern stick.

Break the Train (p. 38)

Working in pairs, students put together two-color pattern trains of cubes and break them down into two-cube "cars"—the repeating units of the pattern.

Make a Train (p. 40)

Students roll a color cube to make two-cube or three-cube "train cars" that they put together in a repeating pattern to make a 12-cube train.

Mathematical Emphasis

- Recognizing a pattern
- Constructing and extending a pattern
- Reading a pattern
- Recording a pattern
- Predicting what comes next in a pattern
- Identifying the unit of a pattern

Teacher Support

Teacher Notes
Reading Patterns (p. 43)
What's the Unit? (p. 45)

Dialogue Boxes
I Think It's Green (p. 31)
A "Harder" Pattern (p. 44)

What to Plan Ahead of Time

Focus Time Materials

Patterns on the Pocket Chart

- Pocket chart for wall display (A Hundred Number Wall Chart is provided with the kindergarten materials kit for this purpose. Since the number cards are not used for this activity, we call it simply a *pocket chart*.)
- Yellow and green construction paper, 1 sheet of each, cut in 2-inch squares and laminated if possible
- What Comes Next? Cards (p. 106): 3 sheets for the class, prepared as indicated on the blackline master
- Small resealable plastic bags: 2 per pair, plus extras
- Color tiles: 1 or 2 sets, sorted by color into 2 plastic bags for each pair of students, one bag containing 7 green and 7 yellow tiles, the other bag 10 each of two other colors
- Small paper cups: 8–10 per pair

Choice Time Materials

What Comes Next?

- Materials for making linear patterns, such as pattern blocks, interlocking cubes, teddy bear counters, keys, shells, buttons, lids, or bottle tops: 1 bin per 4–6 students
- Pattern Paths from Investigation 1: 1 per pair (provided as an option)
- Small paper cups: 6 per pair

Pattern Block Snakes

- *The Sultan's Snakes* by Lorna Turpin (England: Child's Play, 1996) (optional)
- Pattern blocks: 1 bucket per 4–6 students
- Paper pattern blocks, manufactured or copied from masters (pp. 121–126)
- Strips of white construction paper or card stock, 3 by 12 inches: 2 per student, plus extras
- Glue sticks or paste

Add On

- Interlocking cubes: 1 tub per 4–6 students
- Number cubes 0–2, made from blank 1-inch cubes and stick-on labels: 1 cube per pair

Break the Train

- Interlocking cubes: 1 tub per 4–6 students

Make a Train

- Resealable plastic bags: about 6 for the class
- Interlocking cubes: at least 3 sets of 100
- Color cubes, made from blank 1-inch cubes and stick-on labels: 6 cubes for the class
- Color Cars (p. 107): 6 sheets for the class, cut apart and colored as indicated on the blackline master

Patterns on the Pocket Chart

What Happens

The pocket chart is introduced as a tool for exploring what comes next in a pattern. In a whole-class activity, students identify, read, and extend a linear pattern of colors on the pocket chart. Then, working in pairs, they each create a linear pattern with color tiles and try to add on to their partner's pattern. Their work focuses on:

■ describing, copying, and extending a pattern

■ predicting what comes next in a pattern

■ identifying the unit of a pattern

Materials and Preparation

■ Post a pocket chart on the wall.

■ Arrange an a-b pattern in the first row of the pocket chart, using green and yellow paper squares (10 cards). Beginning with the fifth pocket, cover the last six colored squares with What Comes Next? cards (prepared from the master on p. 106).

■ For each pair of students, prepare a supply of color tiles in two resealable plastic bags. In one bag, place at least 7 green and 7 yellow color tiles. In the other bag, place at least 20 tiles in two colors (10 of each color).

■ Provide 8–10 small paper cups (no more than 2 inches across the top) for each pair of students.

Patterns on the Pocket Chart

Note: Patterns on the Pocket Chart is suggested for use as an ongoing classroom routine in kindergarten. If you plan to repeat this activity throughout the year as a classroom routine, you can periodically change the colors of the squares. You might also create patterns with pattern blocks or other small objects (cubes, buttons, keys) that will fit into the pockets.

Seat students where everyone can see the pocket chart and distribute a plastic bag of green and yellow tiles to each pair. Since students will be using these tiles to build a pattern, give them room to work, either on the floor in the meeting area or at tables.

Using the tiles in your bag, work with your partner to build the same pattern that you see on the pocket chart. For right now, build only the part of the pattern that you can see. What do you notice about the pattern?

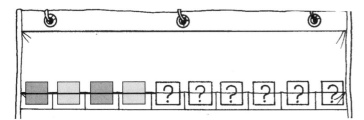

Some students may comment on the green and yellow cards, while others may notice the question marks. Some kindergartners may recognize and know the name for the question mark, but most will not. Discuss the use of the question mark to indicate that a question is being asked.

Each time you see one of these cards with the question mark, I want you to think of this question: "What comes next?" That is, what color might be hidden under the card? See if you can make a prediction, or a guess, about what color is under each question mark on our chart. Use your tiles to show what color would come next. How do you know?

The **Dialogue Box**, I Think It's Green (p. 31), illustrates how one kindergarten class predicted what color came next.

Now work with your partner to add on to this pattern, using your yellow and green tiles.

This approach—asking student pairs to copy, predict, and extend the pattern on the pocket chart—gives you an opportunity to quickly observe their understanding of the activity. It also gives every student a chance to respond, instead of just a single student called on.

When everyone has made a green-yellow arrangement at least five or six tiles long, ask the whole group to "read" the pattern together. Verbalizing the pattern they have built (or the pattern they are considering) often helps students internalize the pattern or recognize an error that might exist in their pattern. This is further discussed in the **Teacher Note**, Reading Patterns (p. 43).

Point to each of your tiles as we read this pattern *[point to each pocket chart square as you read]*: **green, yellow, green, yellow. Now we come to the first question mark. Don't say the next color—just point to the color you think is under this card.**

After you observe the students' responses, uncover the next color by removing the What Comes Next? card. Ask students if their prediction was correct, then continue through the pattern.

What color should the next card be?

Continue in this manner until all the colored squares have been revealed. If students have used more than ten color tiles, assure them that the pattern can indeed continue, and that you just ran out of room on this row of the chart.

Depending on your students, you may want to present another pattern on the pocket chart using yellow and green squares. For example, you might build an a-a-b (yellow-yellow-green) or a-b-b (yellow-green-green) pattern and see if students can extend it using their tiles.

Activity

Patterns with Color Tiles

Now you and your partner are going to make your own patterns with the color tiles. It will be like a game. One person at a time will build a pattern and then hide some of the tiles under these small cups. Each of you will take turns guessing the colors of the tiles that are hidden.

Explain that students should use only two colors in their pattern. Demonstrate making a tile pattern and using the small cups to cover over the last few tiles. Point out that this is a lot like the game you played with them on the pocket chart.

Then ask two student volunteers to demonstrate the same process. While one student arranges a pattern of tiles and covers the last few tiles, the other student hides his or her eyes. When the pattern and the cups are in place, Player 2 uncovers his or her eyes, builds a copy of the pattern that is showing, and then tries to add on to it. (Or, students can just explain what they think comes next and why.)

Some students become very creative with this game and try to be tricky about how they hide tiles. See the **Dialogue Box,** A "Harder" Pattern (p. 44), and **Teacher Note**, What's the Unit? (p. 45) for information about how to turn these situations into "teachable moments."

Explain to students where they will find the tiles and the paper cups needed for this activity. You might distribute a bag of tiles and a set of six small paper cups to each pair as they leave the meeting area, with students then free to choose their own work space in the classroom. Another option is to set out materials on tables around the classroom before the activity and send student pairs to specific tables to work. Or, you might have a central location for storing the materials and make students responsible for getting what they need.

This game could be played during Choice Time or in a whole-group meeting. If you play with the entire class, use the pocket chart with the What Comes Next? cards to demonstrate the pattern, but provide individual bags of tiles so that all students can participate in extending the pattern.

Observing the Students

Consider the following as you watch students playing this game.

- How do students construct patterns with two colors? Do they always make an a-b pattern? If they make red-yellow-red-yellow, do they see that as the same as or different from yellow-red-yellow-red? the same as or different from green-blue-green-blue? Can they explain why?

- Are students branching out and trying a-b-b patterns or a-a-b patterns?

- How do students extend existing patterns? Can they describe how they know what comes next?

- You can increase the challenge of this activity by asking students to predict the color of a hidden tile further along in the pattern. Can they use the information they have to predict the color of *any* tile that is hidden from view?

- Can students show you the basic unit of the pattern? See the **Teacher Note,** What's the Unit? (p. 45), for further explanation.

Focus Time Follow-Up

 **Choice Time**

Five Choices On the first day of Choice Time, you might offer What Comes Next? (p. 32), a natural extension of the students' work with color-tile patterns in Focus Time, and Pattern Block Snakes (p. 34). The next three activities offer work on identifying the unit of a pattern. You might introduce Add On (p. 36) and Break the Train (p. 38) during the second day of Choice Time, and Make a Train (p. 40) on the third day. On subsequent days, students may choose from all five activities.

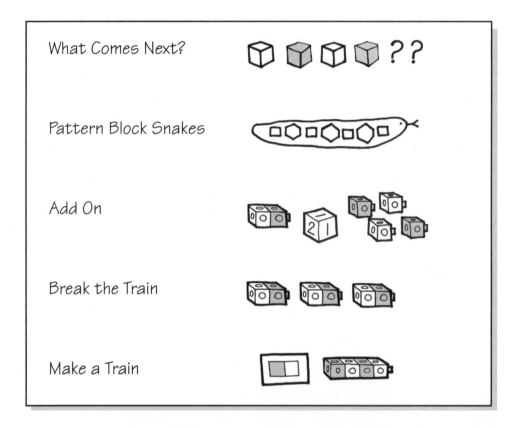

Patterns on the Pocket Chart Many teachers have found that students are very interested in and eager to create their own patterns on the pocket chart using colored squares or other shapes. With this in mind, consider making Patterns on the Pocket Chart an additional option for Choice Time.

I Think It's Green

During Focus Time in Investigation 2, these students are describing the arrangement of colored squares in the pocket chart.

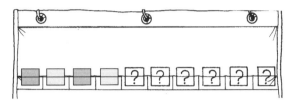

With your tiles, build the pattern that is on the pocket chart. What do you notice?

Henry: *[Points to the question mark]* What's that?

Luke: It's like a fishhook, only it's upside down.

Renata: Look, it's over here too. *[She points to the end mark on a poster that says "How Many Days Have We Been In School?"]*

Ayesha: Yup, it's in books, too.

I think we can find this symbol in lots of places. What do you suppose it means?

Justine: I don't know.

Luke: We could turn it over and pretend to catch something.

Kylie: Maybe we are supposed to guess the next one.

Oscar: Yeah, like a pattern . . . green, yellow, green, yellow, green. Green goes next.

Ayesha: In books it means you have a question. It's called an explanation mark. That's what my mom said when we read.

Gabriela: My dad said it's like when I ask a lot of questions.

You all have lots of ideas. It is used in books, and it does mean you are asking a question. It is called a *question mark*. When Kylie said that maybe we are supposed to guess a color, that is what I was thinking when I put the question mark cards in the pocket chart. When Luke said it looks like an upside-down fishhook, I realized I had never thought of it like that before. It *is* like an upside-down fishhook.

Now see if you can add on to the pattern. Can you make predictions about what color is under each of the question marks? Show me what comes next. How do you know?

Students break into pairs to continue their work.

Ida: I think it will be green. See green, yellow, green, yellow. It has to be.

Tess: I think it will be green, too.

In a second pair, the discussion goes a little differently.

Jacob: I think it will be yellow.

Ravi: No, it has to be green. See *[pointing]*, green, yellow, green, yellow.

Jacob: It doesn't have to be green. It can be yellow. I like yellow. I want there to be two yellows.

This discussion continues back and forth until Ravi goes to the teacher for help.

Ravi: Jacob thinks it is yellow and I think it is green. I know it is green.

You seem to be very sure. Let's see what Jacob thinks. Why do you think it is yellow? Does anyone else think it is yellow?

Jacob: I like yellow. I want there to be more yellows.

Having favorite colors is important, but remember you are trying to predict what comes next in the pattern. What does the pattern tell you so far? Let's read it together.

Everyone: Green, yellow, green, yellow.

Jacob: I guess it will be green. Even though I wish it was yellow.

What makes you think so?

Jacob: Because green comes after yellow.

Choice Time

What Comes Next?

What Happens

Working in pairs, students take turns making patterns with different types of materials. They build their patterns on Pattern Paths. After the pattern-builder hides the last six objects on the path, the partner copies the part of the pattern that is visible and then predicts the hidden pieces. Students' work focuses on:

- constructing, copying, and reading a pattern
- predicting and confirming what comes next
- continuing a pattern

Materials and Preparation

- Set out bins of materials such as pattern blocks, interlocking cubes, teddy bear counters, keys, shells, buttons, lids, or bottle tops (1 bin for each 4–6 students)
- Make available the Pattern Paths (made from p. 105), 1 path per pair.
- Provide six small paper cups for each pair working on this activity.

Activity

What Comes Next? is like the game you played with color tiles. You built a pattern with the tiles and hid the last few tiles under paper cups so your partner could guess what comes next. This time you can make patterns with any of our materials, and you will build your pattern on a Pattern Path. Put one object in each section of the path in some kind of pattern.

Demonstrate how to play this game. Ask a volunteer to hide his or her eyes while you build a pattern on the Pattern Path and cover over the last six objects with small paper cups.

In their eagerness to play the game, students may peek. This can upset the pattern maker. Talk with your students about peeking and ask them to generate good solutions for hiding their eyes while their partner is building a pattern.

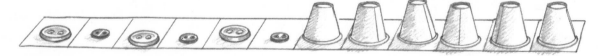

I'm ready, so you can look at my pattern now. The first thing to do is build the part of my pattern that you can see. After that, you can guess what comes next in my pattern. If you are right, I'll lift the cup to show you and you can add that object to the end of your pattern.

Ask your student partner to continue guessing what's next and adding to the duplicate pattern until all the cups have been lifted.

As students play this game in pairs, encourage them to try a variety of materials. Varying the materials makes it more interesting for students to repeat activities over and over. Repetition helps them cement ideas and begin to make connections and generalizations about patterns.

Kindergarten students are eager to guess the next item in a pattern and reveal what is under the cup. You may need to remind them to build their partner's pattern first before they guess what comes next in the sequence.

Observing the Students

Consider the following as you watch students play What Comes Next?

- Can students make a pattern?
- What types of materials do students choose? Do they experiment with different materials or stick solely to one or two?
- What attributes do they focus on when constructing and predicting patterns? For example, do they use color, shape, size, position, quantity?
- Can they "read" their pattern?
- What strategies are they using to predict what comes next?
- Are they able to identify the unit of a pattern?

Some students are likely to construct what they believe to be a pattern but really is not. This situation can cause difficulty for both the pattern maker and the student who is trying to guess what comes next. As students are working, be particularly aware of those who may have difficulty making a pattern that repeats in a predictable way. You might choose to be a partner with such a student, deciding on a pattern together.

The problem of how to avoid a nonpattern would be an excellent topic for group discussion. Involving students in brainstorming solutions to the situation alerts them to the idea that this may happen. Being prepared, they may also be more tolerant if the situation does arise.

Pattern Block Snakes

What Happens

Students use pattern blocks to make repeating pattern "snakes." They record their patterns by gluing paper cutouts of the pattern block shapes onto paper strips. Their work focuses on:

- creating and extending a pattern
- determining what comes next in a pattern
- recording a pattern

Materials and Preparation

- If possible, obtain a copy of *The Sultan's Snakes* by Lorna Turpin (Swindon, England: Child's Play, 1996).
- Make pattern blocks available, 1 bucket for each 4–6 students.
- Provide paper pattern blocks (manufactured or made from pp. 121–126) and glue sticks or paste. **Note:** Students tend to use *lots* of hexagons.
- Provide at least two paper strips, 3 by 12 inches, for each student.

Activity

If you can get a copy of *The Sultan's Snakes* by Lorna Turpin, read it to the class. While this story is not specifically about patterns, each of the snakes pictured displays a different type of pattern. If you do not have the book, simply introduce the activity with a demonstration.

The Sultan had some very interesting snakes. Look closely at this picture. What do you notice about his snakes? How do you think the Sultan could tell his snakes apart?

In this activity, you will use pattern blocks to make snakes. Your snake should have a pattern from one end to the other.

Limit students to two or three shapes per snake. Students will use about 10 to 12 blocks, or enough to cover a paper strip from one end to the other.

Ask a few students to demonstrate patterns they might make. Encourage them to figure out a way for each pattern block to connect to, or touch, the previous block. Some shapes fit together more easily than others.

After you build a snake with the blocks, you will use these paper shapes and a strip of paper to make a paper snake in the same pattern.

Demonstrate how to glue down a paper shape for each block in the pattern. Students will have a variety of techniques for copying their designs. Some will build with blocks on top of the paper strip, then remove one block at a time and glue down a paper shape in its place. Other students will place their paper strip directly adjacent to their block pattern and use their eye to line up the paper shapes in the same pattern. Some students may never use their block pattern for reference as they create their paper record; they have internalized their pattern and do not need a model for support. Still other students may make one pattern with the blocks, then glue down a different pattern with the paper shapes.

When students have finished their paper snakes, help them post the strips on the wall. Some might want to add a tongue and eyes to one end of their snake. Or, you might connect all the strips to make a single, long snake.

Encourage each student to make two snakes. Beyond this, students can decide if they would like to make more. In one class, students were eager to see if they could make a pattern snake stretch all around the classroom. Creating this pattern snake became a class commitment and project.

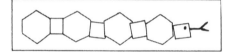 

Observing the Students

With the activity Pattern Block Snakes, students undertake a new task: making a paper record of a concrete model. Consider the following as you observe students.

- What types of patterns are students constructing? Do they choose two or three shapes? How do they arrange them?
- Are students able to continue their patterns? If not, ask them to "read" their pattern to you. Are they able to recognize the inaccuracies when they verbalize their pattern?
- How do students describe their patterns? Do they refer to shapes by color or by shape name?
- How easy is it for students to transfer their block pattern to paper? What strategies do they use?

Variation

Students might use paints or other art materials to design snakes with repeating patterns.

Add On

What Happens

Add On is a game in which students build a two-color pattern stick with 12 cubes. Each player first decides on an a-b unit (blue-red, green-yellow, or the like). Players then build their own pattern stick by rolling a number cube and adding that number of cubes to the stick, to a total of 12 cubes. Their work focuses on:

- constructing the unit of a pattern
- creating and extending a pattern
- counting 12 objects

Materials and Preparation

- Each pair playing the game will need a number cube labeled 0, 1, 1, 1, 2, 2. You can make these with blank cubes and stick-on labels.
- Set out a tub of interlocking cubes for each 4–6 students.

Activity

Introduce this game by pointing out the materials needed to play: a set of interlocking cubes, and a number cube 0–2. Explain that the object of the game is for each player to make a pattern stick with 12 cubes in it. (Although the game could be played individually, playing in pairs keeps the activity moving at a faster pace.)

Each of you first chooses two colors to use in your pattern stick. Snap together two cubes in those colors to start your pattern stick and to remind you what your pattern is. With your partner, take turns rolling the number cube. The number cube will tell you how many cubes you can add on to your stick.

Demonstrate the game. For example, suppose you choose the unit black and yellow. You roll a 2, so you take one black and one yellow cube and add them onto your stick. On your next turn, you roll a 1, so you take one black cube (the next color in your pattern) and add that on after the yellow cube.

Involve students in your turns by asking questions like these:

Right now my pattern is black, yellow, black, yellow, black. Now I have rolled a 2. What cubes should I add to my path?

I want to make a pattern stick with 12 cubes. Do I have enough cubes yet? How many more cubes will I need before I have 12?

When students have 12 cubes on their pattern stick, they put this stick aside, then choose two new colors and play another game.

Encourage partners to check each other's work by verifying the pattern and counting the number of cubes.

Observing the Students

Consider the following as you watch students playing Add On.

- How do students extend their patterns? Do they read the pattern from the first cube every time, or do they look only at the last one or two cubes to know what comes next?

- When the last cube snapped on the stick is not the last cube of the unit (that is, in a red-blue pattern, when the last cube placed is red instead of blue), do students have difficulty adding on the next color of the sequence?

- Are they able to identify the unit of a pattern? Can they figure out how many times the unit of the pattern repeats?

In the *Investigations* curriculum, students are expected to play games like this one many times, over a period of weeks and months. See the **Teacher Note,** Games: The Importance of Playing More Than Once, p. 87. As students grow familiar with a game, introduce some of the variations.

Variations

- Some students may be ready to work with more complex patterns that have units of three cubes in two or three colors. For example: a-b-b (red-blue-blue); a-a-b (red-red-blue), or a-b-c (red-blue-white).

- Players might build a pattern stick with a greater number of cubes.

Break the Train

What Happens

Student pairs take turns putting together two-color pattern trains made with interlocking cubes and breaking them down into two-cube "cars." Their work focuses on:

- constructing patterns
- decomposing patterns
- reading patterns
- identifying the units (repeating parts) of patterns
- counting

Materials and Preparation

- Provide one tub of interlocking cubes for each 4–6 students.
- Snap together a demonstration train using 12 brown and red cubes in an a-b pattern.

Activity

Introduce this activity by showing students your brown and red cube train. Ask the group to count the number of cubes in your train and then read the pattern aloud together, saying each color as you point to the next cube.

What is it that repeats in this pattern? What part goes over and over and over again?

As you gather responses, ask students to explain their reasoning. It may not be obvious to everyone that brown-red is the unit that repeats. To help students see this, break apart the train into six two-cube units of brown and red.

These cubes are like a train. Each car of my train is brown and red. I can make my train longer by adding on more brown and red cars. How many brown and red cars are in my train?

Together with students, count the number of brown and red cars (six). Then snap the cars back together to form one long train.

At Choice Time today, you and a partner can work on making pattern trains. Each of you chooses two colors for the cars of your train, and you can decide how long your train will be. When you have made your train, trade it with your partner. See if your partner can break your train into cars. Then, after you have broken a train into cars, see if you can put it back together again.

Because understanding the unit of a pattern is a challenging idea, we strongly recommend that students begin with a-b patterns. Once students are comfortable with an a-b unit, some may be ready to try a-a-b, a-b-b, or a-b-c units. You can also vary the difficulty by suggesting that students build a longer or shorter train. Ten to 12 cubes is a challenging amount for most kindergarten students to begin with. See the **Teacher Note**, What's the Unit? (p. 45), for more information about how kindergartners build an understanding of the unit of a pattern.

Observing the Students

Consider the following as you watch students work on Break the Train.

- What type of two-color patterns students are constructing? (a-b, a-a-b, a-b-b-a?)

- Are students able to identify the unit of the pattern? Are they able to break apart the pattern train into units?

- Can students reconstruct the train after they have broken it apart?

To determine how accurately students are counting and if they are understanding a unit, ask the following questions individually:

- How many cubes did you use in your pattern train?

- How many cars does your train have? How can you tell?

- How many times did your pattern repeat?

Make a Train

What Happens

Students work cooperatively to build two a-b pattern trains. For each train they choose a "Color Car" that establishes the two-color unit of the a-b pattern. Rolling a color cube, they collect cubes to make the cars, then snap the cars together to build the pattern trains. Their work focuses on:

- identifying the unit of a pattern
- continuing a pattern by adding units
- counting

Materials and Preparation

To prepare Make-a-Train Game Bags for each pair working on this activity:

- Fill resealable plastic bags with 48 cubes, 12 each of four colors. Vary the four colors you put in each bag.
- Each bag also needs a "color cube," a blank wooden or plastic cube labeled on four sides with stick-on labels that show the colors of the cubes in that bag. Label the remaining two sides with stars (free spaces).
- Duplicate and cut apart the Color Cars (p. 107) to make six different cars for each bag. Use markers or crayons to color a different combination on each car.

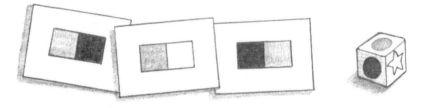

Activity

This activity complements the game Break the Train. As in that game, students are asked to think of the unit of a pattern as one car of a train.

Introduce this activity by showing students the game bags you have prepared. For example, in one game bag, you might have 12 cubes each of red, yellow, green, and blue; a color cube labeled with a yellow dot, a red dot, a green dot, a blue dot, and two stars; and six Color Cars in red-blue, yellow-green, green-red, green-blue, yellow-red, and yellow-blue.

In this game, you and your partner will work together as a team to build two pattern trains. You start by choosing a Color Car for each train.

Ask for a volunteer to help you demonstrate the game. Choose two different Color Cars. Snap together two cubes in those colors to form the first car of each train.

Shanique has chosen blue-green and I have chosen green-yellow. Together we are going to make two trains. One will be a blue-green train, and the other will be a green-yellow train.

We will take turns rolling this color cube. It will tell us what color cube we can take to make our train cars. If we roll a star, we can take any color.

Demonstrate how to roll the cube. The players take one cube on every roll of the color cube. If the color rolled is not in one of the two trains, the same player rolls again.

As we collect cubes of different colors, we snap them together in twos to make cars for our trains. We can't snap just one cube onto a train—we need to have a whole car first.

Be sure students understand that they cannot add cubes onto a train until a car is complete, with both colors. For example, for a blue-green pattern train, players would need both a blue and a green cube before the two could be snapped, as a unit, onto the train. Some teachers have found it helpful to give players an index card to use as a "depot" where they can place loose cubes as they acquire them on each roll, until they have two cubes that make a car.

You may need to remind students that they are working *together* to make two trains, so they need to consider the pattern colors of both trains. The goal is to make two trains that are 12 cubes long. When players think they have completed a train, they should count to be sure they have 12 cubes, and double-check that their pattern is correct.

Observing the Students

Consider the following as you watch students playing Make a Train.

- Are students able to replicate the units on the Color Cars with cubes?
- Are students able to plan ahead and think about how to use cubes when they cannot complete a Color Car on that turn? When and how do they add units onto their trains?
- How do players use the star when they roll it?

When students have successfully built two trains, they should break them apart and place everything back into their Make-a-Train Game Bag (the interlocking cubes, the color cube, and the Color Cars).

Variation

Once students are comfortable with the basic game, you can add some three-cube Color Cars to the bags, colored in two or three colors to make a-b-b, a-a-b, or a-b-c patterns.

Reading Patterns

As 5- and 6-year-olds construct patterns, they begin to see similarities and differences in the patterns they make. It is common to hear observations like these:

"That's just like the other one. This one is red-white-red-white, and that one is blue-green-blue-green, but they are the same."

"That one goes black-blue-black-blue, and mine is blue-black-blue-black. It's the opposite"

These statements of similarities and differences are generalizations about the structure of patterns.

Asking students to "read" their patterns aloud helps cement their understanding of a pattern sequence. Also, verbalizing what they see often gives students clues about what comes next. Reading their pattern in more than one way helps in forming generalizations. For example, students can read the following pattern by color (yellow-orange-yellow-orange) or by shape (hexagon-square-hexagon-square) and see that these are two different ways of describing the same pattern.

Over time and with experience, students will begin to consider how this pattern is the same as and different from an alternating pattern of blue rhombus, green triangle.

In time students can be encouraged to generalize about patterns by identifying the type of *unit* that is used in a repeating pattern. One common convention uses letters to label types of repeating patterns. For example an a-b-a-b pattern is a repeating pattern with two variables, and the unit that repeats is a-b. This is one way to establish a vocabulary of pattern types.

While mathematicians use these labels, you will not necessarily want to use them with your kindergarten students. When we use letters in this way, we are making a generalization about the order of things in the pattern. Using letters to designate the order of things can be confusing to children who are just beginning to associate letters with sounds and words.

You can certainly model the use of language that labels patterns, just as you do with other mathematical terms. However, generalizing about pattern types is challenging for most 5- and 6-year-olds. When they do generalize, it is likely to be in their own language for describing patterns. Listen to how your students label the patterns they are making. For example, in one classroom students always referred to a-b patterns as "candy cane" patterns. This label derived from a memorable red-and-white pattern that they saw early in this unit.

A "Harder" Pattern

Observing students as they work in pairs constructing and predicting patterns can yield some very interesting discussions. In one Choice Time session, Charlotte and Felipe worked as partners on What Comes Next? Felipe started by making a red-yellow pattern with 12 tiles. He covered the last eight tiles with paper cups. From the four tiles showing (red-yellow-red-yellow), Charlotte was quickly able to guess and match the pattern. Then it was her turn

Charlotte: OK, close your eyes. I'm going to make it harder. *[Using the unit red-yellow-yellow, she lays out an a-b-b pattern with 12 tiles and covers the last 10 with cups, leaving only red, yellow showing.]* I'm ready.

Felipe: That's not hard. *[He puts out red, yellow and continues to build red, yellow, red, yellow until he has 12 tiles in his pattern.]*

Charlotte: *[giggling]* Nope!

Felipe: It *has* to be.

Charlotte: No, it doesn't.

Felipe: Yes, it does.

[The students continue to argue as the teacher approaches.] **What's going on?**

Felipe: She made a red-yellow pattern so I think red goes next. See, I made the pattern, too.

Charlotte: But that's not right.

Felipe: It has to be!

Let me look more closely at the work both of you have done. This is Charlotte's pattern? *[Points to the path where 10 cups cover the tiles.]* **And this is Felipe's prediction?** *[The students agree.]* **Let's go back a step. What goes here?** *[Points to the first cup.]*

Felipe: I think it will be red. She has red-yellow, so I made a red-yellow pattern.

Charlotte, show us what you have next in your pattern.

Charlotte: OK *[removing the cup]*. See, it's yellow,

not red. *[Felipe looks very frustrated.]*

Felipe, now you can see more of Charlotte's pattern. What do you think now?

Felipe: *[after a long pause]* I get it.

[Felipe begins to remove color tiles from the path he built, until he has only red-yellow-yellow.]

What do you think comes next?

Felipe: I don't know.

Charlotte, show the next piece in your pattern.

[This continues until six pieces of the pattern are revealed: red-yellow-yellow, red-yellow-yellow.]

Felipe: I get it now. *[He lays out six more tiles, making the red-yellow-yellow pattern.]*

Are you sure?

Felipe: I think so.

Charlotte, show us the rest of your pattern and see if Felipe's matches now.

Charlotte: It does. See, I told you I was going to make it hard!

Yes, you did. What made it so hard?

Charlotte: I made a pattern with one red and two yellows.

Yes, this kind of pattern is tricky, but I see another reason why it was so hard. Felipe, why was this one so hard?

Felipe: It wasn't fair. When it was just red and yellow, *anything* could go next.

It did not seem fair to you. Do you think the way Charlotte showed you only a little piece of the pattern is what made it so hard?

Felipe: Yes!

Predicting what comes next in a pattern is hard work. You need to have enough information so you can be sure about what comes next. Can I show this to the class later so we can talk about this? *[Felipe and Charlotte agree.]*

What's the Unit?

Consider this pattern: red-yellow-yellow-red-yellow-yellow-red-yellow-yellow. We can decompose this repeating pattern into the sections that recur. In so doing, we determine the *unit* of the pattern—in this case, red-yellow-yellow.

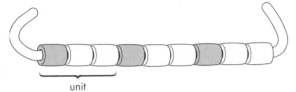

unit

Seeing what elements repeat, or what relationship exists between elements in the pattern, is an important mathematical idea that is challenging for many kindergarten students.

As you observe students in this investigation, take time to talk with them about the part of a pattern that repeats. In the **Dialogue Box**, A "Harder" Pattern (p. 44), Charlotte moves beyond an a-b repeating pattern and tries to make a harder one for her partner Felipe. But when she reveals only two elements of her three-part pattern, she does not give her partner the chance to see the whole unit. Felipe bases his prediction on his prior experience with two-part (a-b) patterns. Once he sees the next piece, he is confused because it does not match his scheme for how patterns repeat.

After being provided with more information, Felipe is able to adjust his thinking and successfully predict the remaining hidden elements of the pattern. Through this interaction, the teacher is able to focus attention on several issues, including clearly identifying the unit that repeats, and the importance of having enough information in order to accurately predict what comes next.

In a similar situation, another student sets up a pattern like this, with only red tiles showing:

By covering up every other tile, he makes it impossible for his partner to predict the hidden pieces. There is not enough evidence to form a solid prediction because the unit of the pattern is not apparent. In situations like these, students often think they are being "tricky" when in fact they are failing to realize some very important ideas about patterns. Make the most of occasions like these to help students clarify their thinking.

In the activity Break the Train, students encounter the idea of a pattern unit in a different way as they are asked to decompose pattern "trains" into "train cars," or the units that make up the pattern. For example, one student builds a red-blue-red-blue-red-blue train of 12 cubes, and her partner's task is to break the train apart into six individual red-blue train cars, the repeating units of the pattern.

Initially, when students are working with a-b pattern trains and cars of just two cubes, this task may seem easy. However, as students begin to work with more complex patterns, such as a-a-b or a-b-b, it is not uncommon to see them continue breaking these trains into sections of two cubes rather than into complete units of three cubes. For these students, recognizing the unit of an a-b or even an a-b-c pattern is easier than finding the unt in a-a-b or a-b-b patterns. As they gain more experience and become familiar with more complex patterns, they will also bring to these their understanding of how patterns work.

You may find that you need to offer students multiple experiences with the activities in this investigation as they begin to think about patterns in new ways.

Hopscotch Paths

Focus Time

Hopscotch Paths (p. 48)

In this investigation, students continue to explore repeating linear patterns and to record on paper the patterns they are building. They start by arranging large squares on the floor to generate hopping patterns and "paths" that resemble hopscotch boards. They then make a model or replica of their life-size path by gluing small paper squares onto strips of paper.

Choice Time

Hopscotch Paths (p. 56)

Student pairs continue to generate life-size hopscotch patterns.

Tile Paths (p. 58)

Students make miniature hopscotch paths with color tiles and represent them with paper squares glued on paper strips.

Continuing from Investigation 2

Pattern Block Snakes (p. 34)

Add On (p. 36)

Break the Train (p. 38)

Make a Train (p. 40)

Reminder About this Unit

If you prefer to present this unit in two parts as discussed in the Unit Overview (p. I-10), Investigations 3 and 4 can be used successfully as a short unit later in the school year. In this case, you may need to reintroduce the Choice Time activities continuing from Investigation 2.

Mathematical Emphasis

- Constructing and extending a pattern
- Interpreting a pattern using physical movements
- Recording a pattern
- Representing a physical pattern using materials
- Predicting what comes next in a pattern
- Identifying the unit of a pattern

Teacher Support

Teacher Notes

About Hopscotch Paths (p. 55)

From the Classroom: What to Expect with Hopscotch Paths (p. 60)

A Pattern Museum (p. 63)

Dialogue Box

One Foot, Then Two Feet (p. 62)

What to Plan Ahead of Time

Focus Time Materials

Hopscotch Paths

- Hopscotch Squares: 8-inch squares of card stock backed with pieces of non-skid rug pad, 10–12 per pair or small group (see p. 48 for other options)
- Adding machine tape (2 rolls) or paper strips 2–3 inches wide, cut into 2-foot strips: 1 strip per student
- Construction paper cut into 1-inch squares and sorted by color into resealable plastic bags
- Glue sticks

Choice Time Materials

Hopscotch Paths

- Hopscotch Squares from Focus Time: 10–12 per pair

Tile Paths

- Color tiles: 1 set per 10–12 students, sorted into single-color bins
- 1-inch colored paper squares, sorted by color into resealable plastic bags
- 2-foot paper strips: at least 1 strip per student
- Glue sticks
- Pattern Pockets (construction paper folded in half with ends stapled to make a pocket about 4 inches deep and 18 inches long): 1 per student, optional

Pattern Block Snakes

- Pattern blocks: 1 bucket per 4–6 students
- Paper pattern blocks
- Paper strips, 3 by 12 inches (supply remaining from Investigation 2)

Add On

- Interlocking cubes: 1 tub per 4–6 students
- Number cubes 0 to 2 (from Investigation 2): 1 cube per pair

Break the Train

- Interlocking cubes: 1 tub per 4–6 students

Make a Train

- Make-a-Train Game Bags (from Investigation 2)

Family Connection

- Add to Our Pattern Museum (p. 108): 1 per family, optional

Hopscotch Paths

What Happens

Students arrange 8-inch squares on the floor to generate hopping patterns and "paths" that resemble hopscotch boards. They then create a model or replica of each life-size path by gluing small paper squares onto strips of paper. Their work focuses on:

- arranging squares to create a pattern
- looking at a pattern and creating a hopping or jumping sequence to match the pattern
- making a permanent (paper) record of a pattern

Materials and Preparation

- Make Hopscotch Squares by cutting card stock into 8-inch squares. Prepare 12 for each pair or small group. To prevent squares from slipping, add a small piece of non-skid rug pad to the back of each square (available from carpet stores or department or hardware stores). Alternatively, square linoleum floor tiles or carpet squares are more durable and will not slip when students hop on them, however they are a little heavy for students to work with. If you have enough non-skid rug pad, you can cut squares from this.

- If your Hopscotch Squares are made of card stock, provide masking tape to hold them together in paths. For ease of use, pre-cut the tape in 2-inch pieces and stick these along the edge of a table or shelf. Or, stick short pieces of tape on pencils or straws and give one to each pair of students.

- Cut adding machine tape (or other paper 2–3 inches wide) into 2-foot strips to provide at least 1 strip per student.

- Using a paper cutter, prepare 1-inch paper squares from construction paper. If you do not have a paper cutter, you could duplicate a sheet of 1-inch grid paper (master provided on p. 117) onto sheets of construction paper and cut on the lines. Prepare about 20–30 squares per student. Store in resealable plastic bags, separated by color.

- Provide glue sticks for student use.

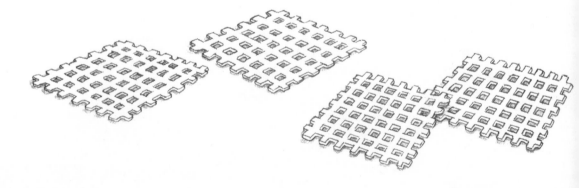

Jumping on a Hopscotch Path

For this activity, students will need ample space to move without disturbing each other. You might seat them either in a semicircle or in two or three rows, facing an open space. A large open space such as a cafeteria or gymnasium would be ideal.

Watch me move. Watch carefully what I do with my feet.

As students watch, hop first on one foot and then on two feet, repeating this pattern as you move across the floor in a simple a-b pattern of movement.

What did you notice about the way I was moving? What did you see me doing with my feet? Who can do what I did?

Call on a few students to take a turn moving across the open space the same way you did. Following the pattern may be difficult for students. Some teachers put words to their actions: "Hop on one foot. Hop on two feet. Hop on one foot. Hop on two feet." However, if you can avoid doing this, you allow students to find their own words later to describe the pattern.

It is important to give every student a chance to try this hopping pattern. You can save time by grouping students into four or five lines, standing one behind the other. At a signal, the first student from each line takes a turn hopping across a designated area.

Have any of you ever played a jumping game called hopscotch? The pattern that you were just practicing is one you might use when you play hopscotch.

While some students may be familiar with hopscotch, knowing the game is not necessary for this activity. See the **Teacher Note**, About Hopscotch Paths (p. 55), for more information.

Once everyone has tried the hopping pattern, show them how to make a "map" of a hopscotch path by placing Hopscotch Squares on the floor in the following pattern:

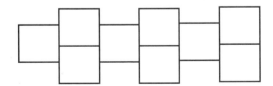

This is a map of a hopscotch path. Do you think you can figure out how to jump just by looking at my map? Can you tell me in words how you should jump on this hopscotch path?

Call on everyone who volunteers to describe the hopscotch path in words. Using words to describe what they see is another way of internalizing the pattern.

As evidenced in the **Dialogue Box**, One Foot, Then Two Feet (p. 62), there are different ways of interpreting this hopscotch path. Some students may say that each square is for one foot, describing the path as "hop on one foot, hop on two feet, hop on one foot, hop on two feet." Other students may interpret the squares as "feet together, feet apart, feet together, feet apart." Either description is fine, and students in your class may even have *other* interpretations. Most important in this activity is that students' descriptions makes sense to them and follow a repeating pattern.

Using another set of Hopscotch Squares, lay out a different path on the floor. If you are using card-stock squares, show students how to use small pieces of tape, about 2 inches long, to tape the path together so that it doesn't slip.

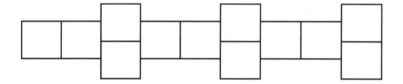

Ask a few students to describe in words how they would hop on this new path. Then have a few volunteers demonstrate how they might hop along the path. As volunteers jump, they or the whole group might describe their movements.

Depending on your class schedule, you can continue now to the next activity, or you might stop here and introduce the next activity another day.

Making Hopscotch Paths

Show students the sets of Hopscotch Squares (12 squares per pair or small group). Explain that now they will have a chance to set up their own hopscotch paths.

Find a place in the room where you can lay out a hopscotch path. Make a pattern with 10 or 12 large squares. See if your feet can follow the pattern.

Indicate the areas of the classroom where students can work. You may need to move furniture temporarily so that pairs have ample space. Be prepared—this will be a very active session. If possible, arrange to use a large open space such as a gymnasium, cafeteria, or the playground.

See the **Teacher Note**, From the Classroom: What to Expect with Hopscotch Paths, for a discussion of the difficulties one class had with this activity and ways the teacher helped the students.

Plan to spend about 10–15 minutes on this activity, or at least long enough for each pair or small group to make and hop along at least one hopscotch path. Give students a 5-minute warning before ending their work. Explain that each pair or group should choose one of the paths they made to share with the class.

Observing the Students

As you watch students making and jumping on their hopscotch paths, consider the following:

- Are they able to make a pattern? If so, what type of pattern do they make? Do they make only a-b patterns, or do they make more complex patterns? If students think they have made a pattern but in fact have not, ask them to explain why they think their arrangement is a pattern. You might point out the part of their pattern that confuses you and see if they can rearrange the squares.
- Can students add on to their pattern?
- How do students describe their path?
- Can students identify the unit of their pattern? That is, can they tell you what part of their pattern repeats?
- Can students figure out how many times the unit of their pattern repeats?

Activity

Sharing Our Hopscotch Paths

I can see a lot of different hopscotch paths. Who would like to show us your path?

The class might move around as a group to view each of the paths. Discuss as many different paths as time allows. Students with similar patterns might share their paths together or one after another. You may find some similar arrangements that students have interpreted in different ways.

I noticed that Xing-Qi and Henry made the same type of pattern with their squares as Alexa and Shanique, but they have different ways of hopping on the same path. Could you show us?

While one student demonstrates how to hop along the path, the other describes aloud what is happening. After both pairs (or groups) have shared their approach, ask the class to compare the two paths.

How are these two hopscotch paths the same, and how are they different?

As you gather responses, you may discover other students who have made a similar pattern but have yet another way of hopping on their path.

Explain that students will be saving their hopscotch paths for another activity. For identification, students write their names on the masking tape holding together the paper squares. This way the squares can be reused after the paths are taken apart. Again, depending on your schedule, you may want to stop here and do the next activity another day. If you plan to break here, store the taped-together hopscotch paths in a safe place.

Note: If students have used carpet, rug pad, or linoleum squares for the hopscotch paths, these are too difficult to save for use on another day. In this case, either plan to continue immediately with the next activity, or plan to allow extra time for Paper Hopscotch Paths on another day so that students can first reconstruct their life-size path with the larger squares.

Paper Hopscotch Paths

Introduce this activity by displaying a hopscotch path of your own or a student-made path.

You and your partner designed a hopscotch path similar to this one. We used the word *pattern* when we talked about our hopscotch paths. How is this path like a pattern?

As students share, listen for statements that suggest they are noticing that part of the path repeats or continues over and over again.

You and your partner are going to work together again with the hopscotch path you saved from earlier. You will start out by finding your path and hopping along it again.

Show examples of the 2-foot paper strips and the 1-inch squares of colored paper.

I will be giving you some small colored paper squares like these, a long strip of paper for each person, and some glue. You will make a small copy of your hopscotch path by gluing these small squares onto the strip of paper. You should choose just one color to work with.

Demonstrate how to copy a larger path with the small paper squares. Using a single color rather than an assortment helps students focus on the *arrangement* of squares. Arrange the small squares along the paper strip in the same pattern and glue them down. When you have finished your copy, ask students to double-check your work to make sure the smaller version matches the larger hopscotch path. Also ask them to extend your path by adding on more paper squares.

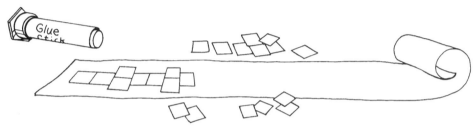

For the rest of the activity time, students work to make their own permanent representation of their large hopscotch path. They may work in pairs or groups, but each student makes his or her own paper copy. Even though their large hopscotch paths may have used only 10–12 large squares, students should extend their smaller paper representations so that they fill the 2-foot paper strip. If students finish early, they can double-check each other's work.

Observing the Students

Consider these questions as you watch students working on Paper Hopscotch Paths.

- Are students able to make a smaller paper version of their life-size hopscotch path? How do they do this? Do they transfer one square at a time? Or do they just seem to know what their pattern is and glue down squares without referring to the original path?
- Are students able to add on squares to extend their pattern?
- Are students able to identify the unit of their pattern that repeats?
- What words do students use to describe their hopscotch path?

Focus Time Follow-Up

 **Extensions**

Hopscotch Around the World As an extension of their work with Hopscotch Paths, you might play the game of hopscotch as a class. The book *Hopscotch Around the World* by Mary Lankford (New York: Morrow Junior Books, 1992) is an excellent resource for rules and variations of how this game is played in different cultures and countries.

Pattern Museum During this unit, students generate many patterns with pattern blocks, color tiles, and interlocking cubes. They also make paper replicas of their patterns. Set up a Pattern Museum to display some of these on a bulletin board, a window ledge, or a table. See the **Teacher Note**, A Pattern Museum (p. 63), for more ideas about organizing this.

 **Homework**

Patterns from Home Send home the letter about Add to Our Pattern Museum (p. 108). Students look for patterns at home and bring in an example of a pattern to display in the museum. If some students have difficulty finding patterns at home, invite them to look around the classroom for objects with patterns.

Choice Time

Six Choices Introduce two new activities that are similar to the Focus Time activities: Hopscotch Paths (p. 56) and Tile Paths (p. 58). On succeeding days, reintroduce four choices from Investigation 2 (or one of their suggested variations): Pattern Block Snakes (p. 34), Add On (p. 36), Break the Train (p. 38), and Make a Train (p. 40).

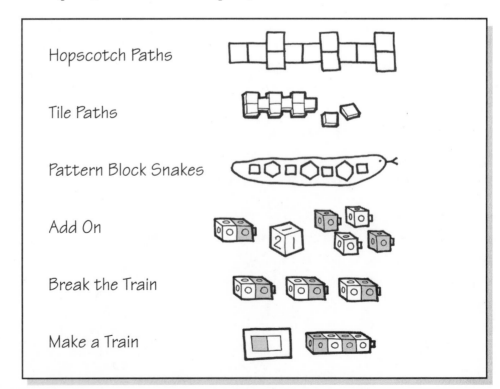

About Hopscotch Paths

Hopscotch is a game played by children all over the world. It has many different variations and is represented in many different ways. This game may be familiar to some of your students but not to others. For the Focus Time activity, Hopscotch Paths, students need not be familiar with the rules of hopscotch since they will not be playing a game but rather generating a hopping pattern and a "path" that resembles hopscotch boards.

Through this activity, students experience constructing, extending, and representing patterns. As they arrange large squares on the floor, students consider which placement of squares will result in a predictable series of hops or footsteps that repeats over and over again. As students move along their hopscotch paths, perhaps hopping on one foot and then on two, their movements create a physical pattern.

Finally, when students make a small-size replica of their larger hopscotch path by gluing down small paper squares on a long paper strip they are making a representation of their pattern—a model that could be used by someone else to recreate the original pattern. By engaging in multiple layers of the same patterning activity, young students build a stronger understanding of what a pattern is and how it is constructed.

Hopscotch Paths

What Happens

Students continue to generate hopping patterns and make hopscotch paths as introduced during Focus Time. Their work focuses on:

- arranging squares to create a pattern
- interpreting a pattern using physical (hopping) movements

Materials and Preparation

- Make available the Hopscotch Squares, 10–12 per pair or small group, from Focus Time.
- If your Hopscotch Squares are card stock, provide masking tape pre-cut in 2-inch pieces to hold the squares together.

Activity

Students will be familiar with this activity from Focus Time (p. 48). In this version, the emphasis is on arranging paths and jumping on them; students will not be making permanent paper models of the paths they arrange.

To introduce this activity, remind students of their previous work with hopscotch paths.

During Choice Time, you and a partner can make more hopscotch paths using our large squares. Arrange the squares on the floor so that they make a pattern you can hop along.

Ask for one or two volunteers to start a hopscotch path on the floor. After two repetitions of the pattern, ask another volunteer to continue this pattern by adding on to the path.

When the path is complete, ask another volunteer to demonstrate one way to hop along the path. After one student demonstrates, ask for another volunteer. Encourage students to think about and test out different hopping patterns for the same path.

Take a close look at the path that Justine has made. Who can figure out one way to jump or hop along this path?

Does anyone have an idea for a different way that you could hop along this same path?

You will probably need to set some guidelines and expectations for student behavior as they are making and jumping on their hopscotch paths. Since this activity requires physical movement and open floor space, it's probably best to limit it to two pairs at a time. If many students are eager to work on this activity, you may also want to set a reasonable time limit for each pair. Reassure students that those who are interested will have at least one opportunity to do this activity.

Observing the Students

Consider the following as you watch students lay out hopscotch paths.

■ Are students able to make a pattern? If so, what type of pattern do they make? Are students' patterns becoming more complex, or do they stick with a familiar pattern?

■ How do students describe their path?

■ Do students have more than one way of hopping along their path?

■ Can students identify the unit of their pattern? Can they figure out how many times the unit of their pattern repeats?

Choice Time

Tile Paths

What Happens

Students make small-size hopscotch patterns with color tiles. They record their work using paper strips and squares, as they did during Focus Time. Their work focuses on:

- constructing a pattern
- recording a pattern
- interpreting a pattern using physical movements

Materials and Preparation

- Provide bins of color tiles, separated by color. You might ask students to do this sorting.
- Continue to make available these materials from Focus Time: bags of 1-inch paper squares, separated by color; 2-foot paper strips for each student; glue sticks.
- As students continue to record patterns on long paper strips, consider making a Pattern Pocket for storing their work. You can make a simple pocket folder that will hold pattern strips by folding a sheet of construction paper (about 9 by 18 inches) in half lengthwise and stapling each end. Students can decorate and write their names on their folders. Store Pattern Pockets in a central location or in individual cubbies or storage bins.

Activity

Briefly introduce this activity by reminding students of the life-size hopscotch paths they made during Focus Time. Also show them an example of a path that one of them made by gluing paper squares on a strip of paper.

Instead of life-size hopscotch paths, in this activity, you will make little hopscotch paths with the color tiles. Just as you did with the large squares, you will arrange the tiles in a pattern to make a hopscotch path. Choose just one color to work with.

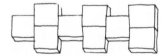

Demonstrate the use of color tiles, placing a few in a hopscotch-style path. Ask a volunteer to extend the pattern by adding on to your tile path.

Once you have a pattern that you like, make a copy of that pattern by gluing paper squares onto a strip of paper. Start your pattern at one end of the paper and continue it all the way to the other end.

As needed, demonstrate or have a student volunteer demonstrate how to make a permanent paper record of the color tile path.

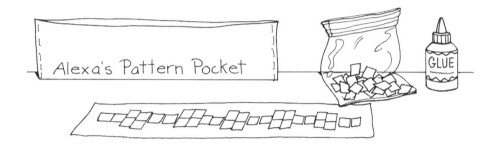

Using your tile path as a model, ask students to describe the pattern using words. Also encourage them to imagine how they would hop along this path if it were made with the large Hopscotch Squares.

How would you describe my tile path to someone? What if you were going to hop along my path—what might the hopping pattern be?

After students have finished making their tile paths and the paper record, they might want to try out the hopping pattern that their path represents.

Observing the Students

Consider the following as you watch students at work on Tile Paths.

- Are student able to make a hopscotch path using color tiles?
- Are students able to make a paper replica of their path using paper squares? How do they do this? Do they transfer one square at a time, or do they just seem to know what their pattern is and glue down squares without referring to their original path?
- Are students able to identify the unit of their pattern that repeats?
- What words do students use to describe the pattern in their hopscotch path?
- Are students able to act out a hopping pattern that matches their tile path?

What to Expect with Hopscotch Paths

The Focus Time activity of Hopscotch Paths unfolded in my classroom over two days. I used our one long day of school (8:30–1:30) to introduce the first three parts of the activity. Then on the following day, which was a regular three-hour day (8:30–11:30), I introduced Paper Hopscotch Paths. I've found that I often need to split a series of Focus Time activities into two or three parts. This works best when I can still schedule the activities close together and not let too much time pass in between.

As with so many of the projects and activities that we do in kindergarten, Hopscotch Paths demonstrated the wide range of abilities that exists within one class of children. Perhaps because this activity also had a physical component, the range seemed even bigger than usual. More than a handful of students had lots of difficulty hopping on one foot. In fact, by the end of the first session, most of my students had adopted the "feet together, feet apart" method of jumping on the path rather than "hop on one foot, hop on two feet."

So, for example, their physical movements for the following path were "Feet together, feet together, feet apart, feet together, feet together, feet apart, feet together, feet together, feet apart."

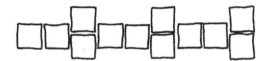

These movements seemed much more reasonable for most of the 5-year-olds in my classroom.

I was not surprised to find that most students, when sent off in pairs to create their own paths, initially copied the path I had used for demonstration. Most were successfully able to jump (or at least approximate jumping) in a pattern sequence when using this path. Designing their *own* Hopscotch Path, however, was challenging for many students.

When they first started laying out their own patterns, two or three paths looked like this:

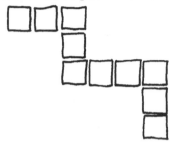

While students were able to hop along the path, there was no pattern or repeating sequence to their plan. They seemed to want to put the squares in a linear pattern and *tell* about their movement, rather than using the position of the square to indicate the expected type of jump. I thought about the shift that students needed to make in this activity. In previous patterning activities, color and shape were the attributes they considered when making and extending a pattern. In this activity, the squares were all one color, so location and position became the important attributes—the ones that indicated how to jump. I could see that they were not all making the needed shift.

After noticing two or three pairs constructing paths without patterns, I decided to call the group back together and focus their thinking by limiting the possibilities. This regrouping seemed pivotal to their experience. The first thing I established was that there were only two choices for laying out the squares: There could either be one square, or there could be two squares placed side by side. I told them:

"When we used the cubes to make a pattern, we chose just two colors to show our pattern. When you make a hopscotch pattern, you also have only a few choices. You can jump on *one foot* or on *two feet*, and you can jump with your *feet together* or your *feet apart*. If you want to jump on one foot or with both feet together, you put one square down on your path. If you want to jump on two feet or with your feet apart, then you put two squares beside each other on your path."

When I explained the two possibilities, a few students remarked "Oh, now I get it."

As I watched pairs of students laying out the large squares, I realized that these were "life-size" manipulatives. While the large squares certainly contributed to the excitement of the activity, I wonder if they also contributed to some of the difficulty students had with making a pattern. With squares of this size, it seemed more difficult for them to have an overall perspective of the path they were making. With the interlocking cubes, they could lay out a design in front of them and really see the whole thing. Squares on the floor were harder to see as a whole pattern.

Over time, though, my students were able to construct more varied patterns using the Hopscotch Squares. Experience was a contributing factor, but I also think that sharing the patterns they made had a large impact on the class. This sharing seemed to broaden their notion of the possibilities for combining the two variables—one square and two squares— into a repeating pattern. In addition, I believe the physical movement of jumping along the path really helped a number of students "see" and experience their path as a pattern.

Below are some examples of the different hopscotch paths students made in my classroom.

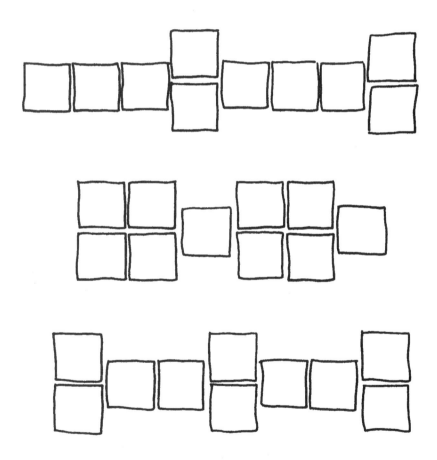

One Foot, Then Two Feet

During Focus Time, these students have been trying out the pattern "hop on one foot, hop on two feet" that the teacher demonstrated. Now the teacher has laid out the hopscotch path using large square tiles, and the class is describing how they might hop along the path.

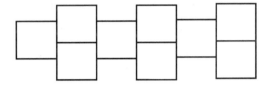

Here's a map of a hopscotch path. How would you describe this map to someone?

Brendan: It's like one square, two squares, one square, two squares.

Miyuki: If you look at it like this *[tilting her head to the left],* then it looks like blocks stacked on top of each other.

Tiana: You could say it goes one, then two, then one, then two.

Carlo: Yeah, it's like a pattern.

Tiana: And it even sounds like a pattern. *[She repeats the one-two pattern in singsong.]*

Jacob: It looks kind of like the tiles on the floor. See, they don't all line up exactly. One is in the middle of the other.

Tiana and Carlo said that my map is like a pattern. How is it like a pattern?

Carlo: Well, it's the same—you know what comes next. After the two at the end, it would be one and then two again, and it would keep going like a pattern.

Tiana: It goes back and forth like a pattern, too.

We were just practicing a pattern by hopping on one foot and then on two feet. Suppose this map was telling you how to jump. Do you think you can figure out how to jump just by looking at my map? Can you tell me in words how you'd jump on this path?

Tarik: First I would hop on one square, and then I would hop on the two squares.

Can you tell us how you would hop? on one foot or on two feet? Would it help to show us?

Tarik: I would hop on two feet like this . . . *[he hops forward on both feet],* and then I would hop and put one foot in each square like this . . . *[he hops forward again, separating his feet].* And I would keep going like this.

So your hopping pattern is about having your feet together or your feet apart. Can you hop for us again like that? *[As Tarik hops across the floor, the teacher verbalizes the pattern]* . . . **Feet together, feet apart, feet together, feet apart. Would someone else like to try Tarik's jumping pattern?**

[As Maddy tries the pattern, the class chants "feet together, feet apart, feet together, feet apart," thus reinforcing the pattern with words.]

Does anyone else have a different way of thinking about this hopscotch path?

Ida: I think it's like we were just doing. First you hop on one foot, then you hop on two feet and then on one foot, and then on two.

So when you hop, you would put one foot in each square.

Ida: Yup.

Oscar: I think hopping on one foot is hard! I like hopping on two feet better.

Does anyone else have a different way of thinking about this hopscotch path? *[There are no other responses.]* **OK, so some people might hop on one foot and then two feet, and other people might hop with feet together and then feet apart.**

You are going to have a chance to make your own hopscotch paths using big squares like these. If you find another way to jump along your path, let me know and you can share it with the class.

A Pattern Museum

A museum is a place for displaying art work and special projects. Some students may have visited a museum before; ask them to tell what they know about museums. Just as there are museums especially for art and other museums for science displays, the class Pattern Museum will be a place to show examples of patterns that students are making or that they find in their environment.

Patterns from School As a class, generate a list of the types of patterns students have been making at school that could be placed in the museum. Students might mention their cube trains or pattern sticks, pattern block snakes, hopscotch paths, and tile paths. Help them look for other examples of patterns around your classroom, such as books that have a repeating pattern in the text or pictures, a design from a poster, a calendar, or student art work.

Patterns from Home Looking for examples of patterns outside of school can be a nice link between home and school and can involve families in looking for and talking about the patterns they see in the world. Before introducing this homework, gather a few examples of objects with patterns on them to share with the class. You might include an article of clothing (such as a striped shirt), a piece of wrapping paper or wallpaper, or a checkered dish towel. Explain that you will be sending home a note to families (Add to Our Pattern Museum), asking them to help the students find examples of patterns for display in the class museum.

Displaying Patterns The Pattern Museum can be on a table or a shelf, or in a special area of the classroom. Find an area large enough so that each student can display at least two pieces of work, one made in school and one from home. Explain what area is being set aside for this museum. Show students how to display their work so it can be viewed easily. Some teachers keep a supply of 3-by-5-inch cards near the display area. As students add a piece of work or an object to the museum, they can write their name on one of these "exhibit cards" and place it with their contribution.

Using the Pattern Museum Once the Pattern Museum has acquired a number of pieces, hold a group discussion about the items on display. You might begin the discussion with one of these questions:

How would you explain the Pattern Museum to someone who was visiting our classroom?

Are all of these patterns? How can you tell?

If we were going to group some of these items together, which ones would you put together? Why?

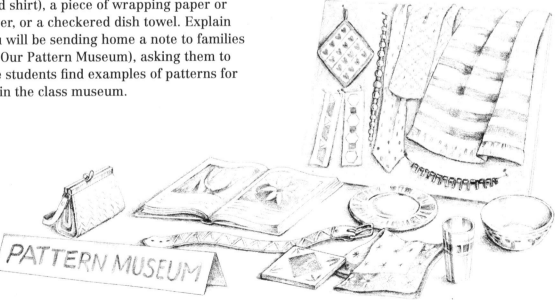

Pattern Borders

Focus Time

Pattern Borders (p. 66)

Students extend their work with patterns on the pocket chart, as introduced in Investigation 2 of this unit. Now they make pattern borders, first by filling in the outer pockets of the pocket chart, then by placing color tiles on square-frame border mats. As they make these borders, students think about what happens to the pattern when it turns a corner. They also notice the relationship between the *last* color and the *first* color placed in their border to determine whether or not their border makes a continuous pattern.

Choice Time

Color Tile Borders (p. 74)

Students build pattern borders with color tiles and record their patterns by coloring border mats.

12 Chips (p. 76)

Students make different two-color linear patterns using 12 chips and record how many of each color they used, then compare the different patterns.

Staircase Patterns (p. 78)

Students place towers of interlocking cubes in "staircase" patterns that go up and back down in a predictable way.

Continuing from Investigation 3

Hopscotch Paths (p. 56)

Tile Paths (p. 58)

Mathematical Emphasis

- Making a linear pattern in a rectangular frame
- Making and comparing patterns that use the same two variables (of color)
- Copying, building, and extending patterns that grow (or shrink) in some regular and predictable way
- Determining a rule for how a pattern grows (or shrinks)
- Recording patterns

Teacher Support

Teacher Note

Why Borders? (p. 73)

Dialogue Box

What Comes Here? (p. 80)

What to Plan Ahead of Time

Focus Time Materials

Pattern Borders

- Examples of borders (e.g., on bulletin boards or posters, or in books)
- Pocket chart for wall display
- Construction paper in two colors, cut in 2-inch squares: 15 squares of each color, laminated if possible
- Color tiles: 2 buckets for the class
- Student Sheets 1–3, Border Mats A, B, and C (pp. 109–111): at least 1 of each per student, plus extras (includes supply for Choice Time)

Choice Time Materials

Color Tile Borders

- Color tiles: 2 buckets for the class
- Border Mats A, B, and C (supply remaining from Focus Time)
- Colored pencils, crayons, or markers
- One-inch grid paper (p. 117): at least 1 per student (optional, for variation)

12 Chips

- Two-color counters (chips with a different color on each side): 12 per student, each set kept in a resealable plastic bag
- Colored pencils, crayons, or markers to match the two-color counters
- Student Sheet 4, 12 Chips in All (p. 112): 1 per student, plus extras
- Pattern Paths (from Investigation 1): 1 per student (optional)

Staircase Patterns

- Interlocking cubes (about 30 per student)
- Staircase Cards A–C (p. 113–115): 4 copies of each to share
- Staircase Grid (p. 116): 3–4 per student, plus extras
- Tape (to tape grids together as needed)

Hopscotch Paths

- Hopscotch Squares (from Investigation 3): 10–12 per pair

Tile Paths

- Color tiles: 1 set per 10–12 students, sorted into single-color bins
- 1-inch colored paper squares, sorted by color into resealable plastic bags
- 2-foot paper strips: at least 1 strip per student
- Glue sticks

Pattern Borders

What Happens

Together as a class, students create a two-color pattern border on the pocket chart. They discuss how the pattern repeats and notice that in this example, the pattern is continuous. Students then make their own pattern borders by placing color tiles around the square frame of border mats of different sizes. Their work focuses on:

■ making and extending a pattern

■ making a linear pattern in a rectangular frame

■ making different patterns using only two colors

Materials and Preparation

■ Find one or two examples of a border to share: on a bulletin board, a poster, or an illustration in a book. Choose a border that goes all the way around an object, rather than along only one or two sides. It is not necessary to find a border with a repeating pattern. Many children's books have beautiful borders around the illustrations; for some examples, refer to the list of Related Children's Literature (p. I-15).

■ Cut construction paper in two colors into 2-inch paper squares, making 15 squares of each color. The example in Focus Time mentions yellow and red, but any two contrasting colors will work.

■ Fill the top row of the pocket chart with paper squares of two colors in an a-b pattern (for example, yellow-red-yellow-red . . .). Keep the remaining 20 squares at hand.

■ Divide two buckets of color tiles into smaller containers for use by pairs or small groups. Pairs will need up to 12 tiles in each of two colors.

■ Duplicate Student Sheets 1–3, Border Mats A, B, and C (pp. 109–111), to provide at least 1 of each per student (includes a supply for Choice Time).

A Border on the Pocket Chart

Show the example of a border on a poster or in a book illustration.

A *border* is a line or a decoration that goes along the edge of something. Sometimes a border goes along only one or two sides of an object. Other times it goes all the way around the object. What do you notice about this border?

After a few students make observations about the border you have displayed, introduce the pocket chart.

Today we are going to build a border around the pocket chart. What do you notice about the border that I have started on the top edge?

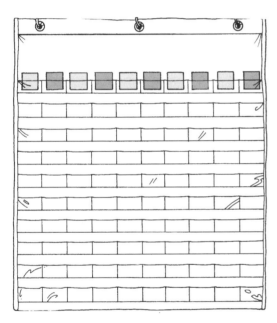

We are going to make a border that goes all the way around the outside edge of the pocket chart. In order to make this pattern turn the corner and continue along this edge *[run your finger down the right-hand edge of the chart]*, what color square do you think would go here? *[Point to the last square in the second row.]* How can you tell?

As you collect responses, use the colored paper squares (in this example, yellow and red) to fill in the pockets down the right-hand column. Occasionally ask, "What color goes here?" as you work your way down the chart. Each time a prediction is made, ask students to explain their thinking before you place that color square in the pocket.

If a student suggests an incorrect color, go ahead and add it to the chart; once students have seen their suggestion, they will nearly always self-correct either themselves or each other. Going back to the beginning and reading the pattern from the start helps many children confirm what color goes next in the pattern.

When the lower right corner square has been filled, return to the first square and read the pattern through one more time.

Do we still have a pattern? How can you tell? What happens if we turn this corner and keep going? Can we continue our pattern?

Continue as before, asking one student at a time to tell you the next color in the pattern border as you fill in the pockets along the bottom of the chart. Stop when you have filled the lower left corner square.

What happens if we turn this corner and keep going? Can we continue our pattern? What do you think will happen when we reach the top of the chart?

Again, continue as before, but as you begin to approach the top of the chart, leave three or four pockets empty and ask students to predict what color square should go in the last empty space.

Being able to predict what comes next is an important idea in learning about patterns, as is being able to predict "what comes here?" in a pattern sequence when not all the intervening information is given. Some students may have a strategy for visualizing the color of the last square along the border. As you collect responses, ask students to explain how they predicted that color. See the **Dialogue Box,** What Comes Here? (p. 80), for an example of this discussion.

Fill in the remaining colors along the left-hand side of the pocket chart to check students' predictions.

We continued our pattern all around the edge of this pocket chart. *[Run your finger around the edge of the chart.]* **We covered the border of the chart with a pattern, so we made a pattern border. There is something special about this pattern border.**

Call students' attention to the upper left corner square. If they do not notice, point out that the color squares of this a-b pattern meet in exactly the right place. In other words, the last square of the border (red, in our example) is followed by yellow, which is the first square in the border, and the first color of the yellow-red pattern.

In this pattern border, the colors of the pattern meet in the right way in the corner so that the pattern keeps going around the border of the chart: yellow, red, yellow, red. We started here *[point to upper left corner]* **and went all the way around the border. Right at this corner, the last piece fit in to keep the pattern going. Even if I turned the corner again, our pattern would keep going.**

Students are often delighted by this visual presentation on the pocket chart. They may say it looks like a picture frame or a square, or they may comment on how pretty the pattern is. Take time to enjoy the aesthetics of this work. Patterns are beautiful to look at.

Not all border patterns will finish exactly in the correct color sequence. This depends on the type of pattern (the number of variables in it) and the size of the border. For example, if a student colors Border Mat B in an a-b-b pattern (as shown at right), the pattern is not continuous. The **Teacher Note,** Why Borders? (p. 73), suggests some ways to think about this relationship.

Pattern Borders with Color Tiles

To demonstrate this next activity, gather students in a circle on the floor. Have copies of Student Sheets 1–3 to show students the three border mats they may choose from.

You and a partner are going to design your own pattern borders, using color tiles and one of these border mats. There are different sizes to choose from. You and your partner first decide which border you want to make, then choose two colors. Use tiles in those colors to make a pattern that goes all around the edge.

Demonstrate how to use the color tiles to fill in one of the border mats. As you did on the pocket chart, use an a-b pattern. After choosing two colors, volunteers can take turns adding a color tile onto the pattern border. With the help of your students, complete one border to show what a finished mat will look like. Occasionally ask students what color they think comes next in the pattern and why. As you approach the final square, ask them to predict the color of the tile and explain their reasoning.

When the border is complete, ask students about the continuity of the pattern as it turns the final corner.

Did the tiles of our pattern meet in such a way that the pattern keeps going around the border? Do you think this will always happen? After you have finished making your pattern border, check to see whether the colors of your pattern keep going, all the way around the border again and again, in the order of your pattern.

Explain where students can get the border mats and the color tiles, and send them off to work in pairs on their pattern borders. Circulate to observe as they work.

Observing the Students

Consider the following as you watch students work on their pattern borders.

- Are students using an a-b pattern?
- Are they able to turn each corner of the border and continue the pattern in a new direction?
- Can students predict what color comes next in their pattern? If you point to an empty square on their border, can they predict what color that square will be? How do they predict this color?
- Can students double-check their pattern to see if it is correct? If they have made an error, can they find what is wrong and fix it?
- Have students found any patterns that do not make a continuous pattern around the border? If so, what is special or different about this pattern?

Patterns made with individual tiles are difficult to bring back to the whole group for sharing. For this reason, plan about 5 minutes at the end of the work period for students to look at each other's designs. Before sharing, students clean up by putting away any color tiles or border mats that are not needed. Be sure they do not put away any pattern borders they have completed.

Students might look at the work of classmates at their table or nearby, or they could walk around the room to look at everyone's work.

Pattern borders made with color tiles are difficult to save. Explain that during the upcoming Choice Time, students will have a chance to build more pattern borders and make a permanent paper record of them.

Focus Time Follow-Up

 **Extensions**

Changing Borders on the Pocket Chart Every few days, make a new pattern border on the pocket chart. After starting a pattern on the chart, ask students to help you fill in the border. Begin with a-b patterns, then gradually vary the pattern type while still using only two colors. Try patterns such as a-a-b, a-a-b-b, a-a-b-a, or a-b-b. Notice which types make a continuous pattern and which do not.

Students at the Pocket Chart Hang the pocket chart in a location where students can reach it. During free time or Choice Time, two or three students at a time can make their own pattern on the pocket chart, using the sets of paper squares or the color tiles.

 **Choice Time**

Five Choices During Choice Time, students continue with Color Tile Borders (p. 74), now recording their borders by coloring the border mat. On the first day of Choice Time, you might offer this and the two familiar activities from the previous investigation, Hopscotch Paths (p. 56) and Tile Paths (p. 58). On succeeding days, introduce the two new independent pattern activities, 12 Chips (p. 76) and Staircase Patterns (p. 78).

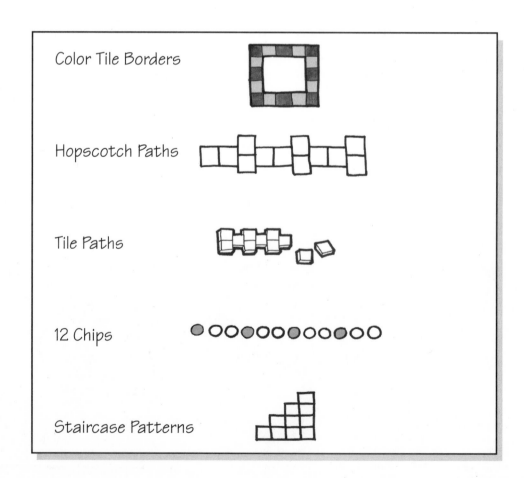

Why Borders?

As students become aware of linear patterns and how they are constructed, they can be encouraged to look at more complex patterns. In addition, it is important to provide different models and examples of linear patterns. The Border Patterns activity provides a nice link between linear patterns and geometric patterns. While a border pattern is technically a linear pattern, the fact that it does not extend in one straight line, but instead forms a rectangular enclosure, allows for some interesting geometric patterns to emerge. Looking at borders highlights both horizontal (left to right) and vertical (up and down) patterns. Asking students to focus on both rows and columns opens up new ways of thinking about patterns.

The corner where the border meets can be a challenging spot. Is the border a continuous pattern or not? The possibilities for making borders seem endless. With the border mats provided, any a-b pattern students make will be continuous, as long as they match one tile to each square. This is because of the relationship between the number of elements in the unit (two) and the total number of squares in any size rectangular border.

As an interesting exercise for teachers (but not kindergartners), investigate which types of patterns make continuous borders and which do not. You can have fun with this, and it illuminates some interesting mathematics. Why do some patterns work and not others? Look at the number of tiles in the border and think about how many times a unit needs to be repeated to make a continuous pattern. Using the pocket chart and a pattern unit of red, orange, yellow, green, blue, indigo, and violet, would this make a continuous border pattern? Why or why not?

Color Tile Borders

What Happens

Students use color tiles to build pattern borders and then record them by coloring the pattern on a border mat. Their work focuses on:

- making a linear pattern in a rectangular frame
- recording a pattern

Materials and Preparation

- Continue to make available from Focus Time the tubs of color tiles and Border Mats A, B, and C (Student Sheets 1–3).
- Students will need colored pencils, markers, or crayons to color the border mats.

Activity

Students will be familiar with this activity from Focus Time. The only difference is that now they get to make a permanent record of their pattern by coloring the border mats.

Start by building several different borders on the border mats. Then choose one that you would like to save and make a record of it by coloring in the squares on the border mat to match your color-tile pattern.

Briefly demonstrate how to color a pattern, to give students an idea of what their record will look like.

It is not uncommon for students to make mistakes in the recording process, either because their original tile pattern does not follow the intended sequence, or because they get confused about which square they are coloring. Because mistakes can cause great concern, acknowledge this likelihood ahead of time and help students think about how they can "fix" any coloring mistakes they make. One idea is to keep a supply of white stick-on labels, cut 1-inch square, to cover incorrectly colored squares.

At this point in the unit, many students may be ready for and interested in making more complex patterns. If they have not already begun to do this, suggest that they make a pattern using two colors *not* arranged in an a-b sequence, such as an a-a-b or a-a-b-b pattern. Ask them to investigate which types of patterns make a continuous border pattern. If appropriate, some students might add a third color to their patterns.

If you have set up a Pattern Museum, suggest that students display one of their favorite pattern borders. (See the **Teacher Note,** A Pattern Museum, p. 63, for more information.)

Observing the Students

Consider these questions as you watch students at work on Color Tile Borders.

- What type of pattern do students construct? Are they beginning to use more complex patterns in addition to a-b patterns?

- Are students able to turn each corner of the border and continue the pattern in a new direction?

- Can students predict what color tile comes next in their pattern? If you point to an empty square on their border, can they predict what color that square will be? How do they predict this color?

- Can students double-check their pattern to see if it is correct? If they have made an error, can they find what is wrong and fix it?

- Can students transfer their work and record it? How do they do this? Do they remove one square at a time and color in that square? Do they get a new border mat and place it next to their tile border, using the original border as a reference? Do some students seem to internalize their pattern and just color it in without referring to the original tile border?

Variations

- Offer students other materials for making pattern borders (pattern blocks, teddy bear counters, interlocking cubes) and for recording their borders (rubber stamps, 1-inch paper squares in assorted colors, and paper pattern blocks).

- Vary the type of paper that students use as a base for their border patterns. Cut squares and rectangles of various sizes, with dimensions in whole inches. (For example, cut standard letter-size paper down to 8 by 11 inches.) You can use unlined paper, construction paper, or 1-inch grid paper for this.

- Some students might be interested in exploring what happens when they make a pattern border around a nonrectangular shape, such as a circle or an oval.

- Ask students to *count* the number of pieces in a completed border. They can count both the total number used and how many of each color or shape were used.

12 Chips

What Happens

Students make different patterns using a total of 12 red and yellow chips. After they build the pattern, they record how many red chips and how many yellow chips they used. Their work focuses on:

- making different patterns with only two colors
- comparing different patterns by looking at the total number of each color used
- considering the unit of a pattern
- counting a set of objects

Materials and Preparation

- Provide two-color counters (chips that are a different color on each side), 12 per student, each set in a resealable plastic bag.
- Duplicate Student Sheet 4, 12 Chips in All (p. 112), to provide 1 per student plus extras.
- Make available Pattern Paths from Investigation 1 for optional use.
- Students will need colored pencils, markers, or crayons of the same colors as their two-color counters.

Activity

Gather students in a circle on the floor as you demonstrate this activity. Alternatively, cut large circles from two colors of construction paper (12 of each color) and tape them to the board or wall for your demonstration. Show a set of 12 two-color counters. Ask students to suggest a possible pattern using the two colors, then build that pattern using all 12 chips. For demonstration purposes you might use a Pattern Path, placing one counter in each square.

Together as a class, read the pattern aloud. Ask students to confirm that it is a pattern. Then focus on the number of chips and colors in the pattern.

Look closely at this pattern. How many red chips are there? (6 red) **How many yellow chips?** (6 yellow) **How many chips altogether?** (12) *[Record this information on the board.]*

We just made a red-yellow-red-yellow pattern. Do you think we can make a different pattern with these red and yellow chips? Who has an idea?

Collect ideas for patterns and demonstrate one. Read the pattern together as a class; then ask a volunteer to count the red and yellow chips. Record this information on your chart under the first set of data. Ask students to confirm how many chips there are in all. This may be obvious to some, but not all kindergartners. Many will need to recount the total number of chips each time they make a new pattern.

We have made two different patterns using this set of chips. Do you think there are other patterns we could make with this same set?

Explain that in this activity, students will look for as many different ways as they can find to make patterns using a set of 12 two-color chips. Demonstrate how students can use Student Sheet 4, 12 Chips in All, to record their work. They color in one row of circles to match each pattern they make, then write the number of each color and the total used.

Students work in pairs for the first part of the activity, reading their patterns to each other to confirm that they have in fact made a pattern. They record their work individually on Student Sheet 4.

Note: Because the circles on the student sheet are necessarily smaller than the actual chips, some students may have difficulty matching their pattern to the circles and will need extra help recording their patterns.

Observing the Students

Consider the following as you watch students working on 12 Chips.

- Are students able to make a variety of patterns using two colors (a-b-b, a-a-b, a-a-b-b, a-a-a-b), or can they make only an a-b pattern? If students are having difficulty with more complex patterns, are they able to extend a pattern such as a-b-b that you start for them?
- Can students count the number of each color used in their pattern? Do they know there are 12 chips total, or do they have to recount the total number each time? Are students noticing any familiar number combinations, such as 6 + 6 = 12?
- Are students able to record their work accurately? Do they use their recording sheet to compare different patterns made with two colors?
- Are students able to make any generalizations about their work? Do they see that a red-yellow and a yellow-red pattern are the same type?

Staircase Patterns

What Happens

Students put together towers of interlocking cubes to make "staircase" patterns and then record these patterns on grid paper. They count the number of cubes in each stair and think about the difference between one step and the next. Their work focuses on:

- copying, extending, and recording patterns that grow (or shrink) in a regular way
- determining a rule for how a staircase pattern grows (or shrinks)
- counting the number of cubes in a step
- comparing the number of cubes in one step to the number in the next step

Materials and Preparation

- Provide interlocking cubes, 1 tub per 4–6 students.
- Duplicate four sets of the Staircase Cards A–C (pp. 113–115).
- Duplicate the Staircase Grid (p. 116), 3–4 per student and some extras. Have tape available to tape sheets together as needed.

Activity

Show Staircase Card A and ask students to describe what they see. They are likely to comment on how the design looks (like steps going up) or on the number of cubes in each step (first there is one, then two, then three). Ask a volunteer to use interlocking cubes to build what is on the card. Together as a class, count the number of cubes in each tower.

Some people think this looks like steps that are going up. In fact, these designs are called *staircases*. If we wanted to make our staircase go up one more step, how many cubes would be in that next step?

As students offer ideas, ask them to explain how they know. Some might say that the next step should have *one more* than the last step. Others might refer to the numbers of cubes, saying that if the last step was 4, then the next step would be 5. Again ask a volunteer to build the next step with interlocking cubes. Place that step next to the other four.

Some people might say this is like a pattern. Do you agree with them or disagree?

Not everyone will see this growing staircase as a pattern. For many students, especially those who are just internalizing what a pattern is, this new kind does not fit their familiar definition of a pattern as something that repeats. Some students, however, will be able to identify the growing pattern in this staircase, with each step increasing by one cube.

Suppose we wanted to make this staircase go back down. What step would come next if you want to go back down the stairs? Tell me what makes you think so.

Ask volunteers to build the next two or three steps back down. Then ask students to look at the *number* of cubes in each step and tell what they notice. Show Staircase Cards A–C and the Staircase Grid.

During Choice Time, you can try to build these three different staircase patterns with the interlocking cubes. Build what you see on the card. Make the staircase go up one more step and then come all the way back down. After you have made your staircase with cubes, record what you did on this Staircase Grid.

Using the staircase you have just built, demonstrate how to color in the staircase on the Staircase Grid. On the lines at the bottom, under each step, write the number of cubes used for that step. When students need a longer grid for Staircase B, tape two sheets together for them.

Observing the Students

Consider the following as you watch students work.

■ Are students able to copy each pictured staircase accurately with cubes? Can they extend the pattern, going up and back down? Can they predict what step comes next in the staircase?

■ How do students describe the staircases? with words? with numbers? Do they recognize the pattern of the staircase? If so, how do they describe that?

■ Can students record their staircases on the grid? Are they able to record the number of cubes in each step? If students are unsure about how to write the numeral but they know how many cubes, suggest that they draw dots or tally marks to show how many.

Variation

Students can make up their own staircase patterns to build with cubes and record on the Staircase Grid. Add these new patterns to the Staircase Cards for other students to copy and build.

What Comes Here?

During Focus Time in Investigation 4, Pattern Borders, this group of kindergartners has just been introduced to making pattern borders on the pocket chart. They have been predicting what comes next in the pattern as their teacher adds color cards to the chart, one by one, making a blue-red (a-b) pattern border around the outside edge. Now the teacher skips ahead and places a What Comes Next? card (the question mark) in the last square of the border.

See where I've put this question mark? This time, instead of telling me what color comes after the last one I put in, I'd like you to think about this pocket where the question mark is. What color will that be?

Alexa: Umm, I think it's going to be blue. No, red . . . no, blue. No, I'm not sure.

Shanique: I know what color it is. It's red.

Jacob: No, I think it's blue.

Tarik: Well, it's going to be blue or red because those are the colors in the pattern. See? Blue, red, blue, red . . . *[the class chants along].*

Does everyone agree that it will either be blue or red?

Many students: *[laughing]* Yes!

Shanique, can you tell us why you think it might be red?

Shanique: See, it ended on red, and I just in my mind went, blue, red, blue, RED. *[From her seat, she points at the chart as though she is keeping track of the squares.]*

Tarik: That's just what I did, too!

So you thought about the part of the pattern that you could see, and then in your mind kept that pattern going until you got to the last square. Did anyone else figure out the problem in the same way Shanique did? *[About six students raise their hands.]*

Jacob, you thought that the color under the question mark would be blue. Could you tell us why you think it might be blue?

Jacob: I thought blue because, see the card at this end of the row *[the last card in the second row]* is blue, and because the card here *[the red card at the end of the first row]* is red. And the next color after red is blue, so I think it's blue.

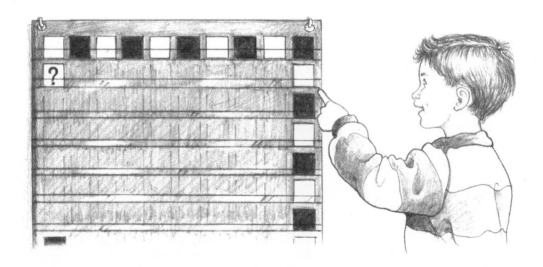

You were looking at the pattern on this [right] side of the chart, and you noticed that blue came after red, and red was at the end of this [first] row, so the card at the beginning of this next row would be blue. Did anyone one else think about the pattern the way Jacob did? *[Two hands go up.]*

Ida: Well, sort of I did. But now I'm not sure it's blue.

Carlo: I don't *think* the card might be red . . . I *know* the card is red! Because I know it in two ways. It does go from red, blue, red, blue, red, but see, there's another pattern. This line [the bottom row] has a red first and a blue last, and the next line has a blue first and a red last, and this line has a red first and a blue last, so this line up here [the second row] has a blue last, so this first one would be red. It's like another pattern.

Carlo sees another pattern on the chart. He sees a pattern about how the rows begin and end. Did anyone else see it the way Carlo did and look at what color the rows started and ended with? *[No one raises a hand, and many students look puzzled.]* **Let's keep that idea in our heads, and the next time we make a border pattern we can test it out.**

Now let's keep adding cards to our border and see what color goes where that question mark is.

[Different students come up and each add a card to the border, which ends with red.]

Alexa: Hey, it *is* red! And look—it matches just right, because blue comes next [the first card at upper left], and it's supposed to come next!

Yes, that's interesting how that happened. We can talk about that idea the next time we make a border design on the pocket chart.

The students in this class used several different strategies to predict the color of the final square of the border design. The most common strategy was to continue around the pattern sequence, past the empty pockets, until they reached the question mark. One student, however, based his prediction on identifying a pattern in the overall chart: Carlo noticed the pattern that the colors make at the beginning and end of rows. This more generalized response is one that some, but probably not many, young students will be able to make about patterns.

The teacher in this classroom offered students another opportunity to explore these ideas by reintroducing a Choice Time activity from Investigation 2, What Comes Next? (p. 32). In this activity, students build patterns with 12 tiles, cubes, or other counters and cover the last six objects with paper cups. A partner who has not seen the pattern then tries to guess what's under the cups. Rather than focusing on the *next* element in the pattern, the teacher now encourages students to guess a hidden element further along, for example, under the fourth cup, or under the last cup. This encourages students to use the information they have to predict that which they cannot see.

Choice Time is an opportunity for students to work on a variety of activities that focus on similar mathematical content. In the kindergarten *Investigations* curriculum, Choice Time is a regular feature that follows each whole-group Focus Time. The activities in Choice Time are not sequential; as students move among them, they continually revisit the important concepts and ideas they are learning in that unit. Many Choice Time activities are designed with the intent that students will work on them more than once. As they play a game a second or third time, use a material over and over again, or solve many similar problems, students are able to refine their strategies, see a variety of approaches, and bring new knowledge to familiar experiences.

Scheduling Choice Time

Scheduling of the suggested Choice Time activities will depend on the structure of your classroom day. Many kindergarten teachers already have some type of "activity time" built into their daily schedule, and the Choice Time activities described in each investigation can easily be presented during these times. Some classrooms have a designated math time once a day or at least three or four times a week. In these cases you might spend one or two math times on a Focus Time activity, followed by five to seven days of Choice Time during math, with students choosing among three or four activities. New activities can be added every few days.

Setting Up the Choices

Many kindergarten teachers set up the Choice Time activities at centers or stations around the room. At each center, students will find the materials needed to complete the activity. Other teachers prefer to keep materials stored in a central location; students then take the materials they need to a designated workplace. In either case, materials should be readily accessible. When choosing an arrangement, you may need to experiment with a few different structures before finding the setup that works best for you and your students.

We suggest that you limit the number of students doing a Choice Time activity at any one time. In many cases, the quantity of materials available establishes the limit. Even if this is not the case, limiting the number is advisable because it gives students the opportunity to work in smaller groups. It also gives them a chance to do some choices more than once.

In the quantity of materials specified for each Choice Time activity, "per pair" refers to the number of students who will be doing that activity at the same time (usually not the entire class). You can plan the actual quantity needed for your class once you decide how many other activities will be available at the same time.

Many kindergarten teachers use some form of chart or Choice Board that tells which activities are available and for how many students. This organizer can be as simple as a list of the activities on chart paper, each activity identified with a little sketch. Ideas for pictures to help identify each different activity are found with the blackline masters for each kindergarten unit.

In some classrooms, teachers make permanent Choice Boards by attaching small hooks or Velcro strips onto a large board or heavy cardboard. The choices are written on individual strips and hung on the board. Next to each choice are additional hooks or Velcro pieces that indicate the number of students who can be working at that activity. Students each have a small name tag that they are responsible for moving around the Choice Board as they proceed from activity to activity.

Introducing New Choices

Choice Time activities are suggested at the end of each Focus Time. Plan to introduce these gradually, over a few days, rather than all at once on the same day. Often two or three of the choices will be familiar to students already, either because they are a direct extension of the Focus Time activity or because they are continuing from a previous investigation. On the first day of

Choice Time, you might begin with the familiar activities and perhaps introduce one new activity. On subsequent days, one or two new activities can be introduced to students as you get them started on their Choice Time work. Most teachers find it both more efficient and more effective to introduce activities to the whole class at once.

Managing Choice Time

During the first weeks of Choice Time, you will need to take an active role in helping students learn the routine, your expectations, and how to plan what they do. We do not recommend organizing students into groups and circulating the groups every 15–20 minutes. For some students, this set time may be too long to spend at an activity; others may have only begun to explore the activity when it's time to move on. Instead, we recommend that you support students in making their own decisions about the activities they do. Making choices, planning their time, and taking responsibility for their own learning are important aspects of the school experience. If some students return to the same activity over

and over again without trying other choices, suggest that they make a different first choice and then do the favorite activity as a second choice.

When a new choice is introduced, many students want to do it first. Initially you will need to give lots of reassurance that every student will have the chance to try each choice.

As students become more familiar with the Choice Time routine and the classroom structure, they will come to trust that activities are available for many days at a time.

For some activities, students will have a "product" to save or share. Some teachers provide folders where students can keep their work for each unit. Other teachers collect students' work in a central spot, then file it in individual student folders. In kindergarten many of the products will not be on tidy sheets of paper. Instead, students will be making constructions out of pattern blocks and interlocking cubes, drawing graphs on large pieces of drawing paper, and creating patterns on long strips of paper.

(Coninued on next page)

For some activities, such as the counting games they play again and again, there may be no actual "product." For this reason, some teachers take photographs or jot down short anecdotal observations to record the work of their kindergarten students.

During the second half of the year, or when students seem very comfortable with Choice Time, you might consider asking them to keep track of the choices they have completed. This can be set up in one of these ways:

- Students each have a blank sheet of paper. When they have completed an activity, they record its name or picture on the paper.

- Post a sheet of lined paper at each station, or a sheet for each choice at the front of the room. At the top of the sheet, write the name of one activity with the corresponding picture. When students have completed an activity, they print their name on the appropriate sheet.

Some teachers keep a date stamp at each station or at the front of the room, making it easy for students to record the date as well. As they complete each choice, students place in a designated spot any work they have done during that activity.

In addition to learning about how to make choices and how to work productively on their own, students should be expected to take responsibility for cleaning up and returning materials to their appropriate storage locations. This requires a certain amount of organization on the part of the teacher—making sure storage bins are clearly labeled, and offering some instruction about how to clean up and how to care for the various materials. Giving students a "5 minutes until cleanup" warning before the end of any Choice Time session allows students to finish what they are working on and prepare for the upcoming transition.

At the end of a Choice Time, spend a few minutes discussing with students what went smoothly, what sorts of issues arose and how they were resolved, and what students enjoyed or found difficult. Encourage students to be involved in the process of finding solutions to problems that come up in the classroom. In doing so, they take some responsibility for their own behavior and become involved with establishing classroom policies.

Observing and Working with Students

During the initial weeks of Choice Time, much of your time will be spent in classroom management, circulating around the room, helping students get settled into activities, and monitoring the process of making choices and moving from one activity to another. Once routines are familiar and well established, however, students will become more independent and responsible for their own work. At this point, you will have time to observe and listen to students while they work. You might plan to meet with individual students, pairs, or small groups that need help; you might focus on students you haven't had a chance to observe before; or you might do individual assessments. The section About Assessment (p. I-8) explains the importance of this type of observation in the kindergarten curriculum and offers some suggestions for recording and using your observations.

Materials as Tools for Learning

Concrete materials are used throughout the *Investigations* curriculum as tools for learning. Students of all ages benefit from being able to use materials to model problems and explain their thinking.

The more available materials are, the more likely students are to use them. Having materials available means that they are readily accessible and that students are allowed to make decisions about which tools to use and when to use them. In much the same way that you choose the best tool to use for certain projects or tasks, students also should be encouraged to think about which material best meets their needs. To store manipulatives where they are easily accessible to the class, many teachers use plastic tubs or shoe boxes arranged on a bookshelf or along a windowsill. This storage can hold pattern blocks, Geoblocks, interlocking cubes, square tiles, counters such as buttons or bread tabs, and paper for student use.

It is important to encourage all students to use materials. If manipulatives are used only when someone is having difficulty, students can get the mistaken idea that using materials is a less sophisticated and less valued way of solving a problem. Encourage students to talk about how they used certain materials. They should see how different people, including the teacher, use a variety of materials in solving the same problem.

Introducing a New Material: Free Exploration

Students need time to explore a new material before using it in structured activities. By freely exploring a material, students will discover many of its important characteristics and will have some understanding of when it might make sense to use it. Although some free exploration should be done during regular math time, many teachers make materials available to students during free times or before or after school. Each new material may present particular issues that you will want to discuss with your students. For example, to head off the natural tendency of some children to make guns with the interlocking cubes, you might establish a rule of "no weapons in the classroom." Some students like to build very tall structures with the Geoblocks. You may want to specify certain places where tall structures can be made—for example, on the floor in a particular corner—so that when they come crashing down, they are contained in that area.

Establishing Routines for Using Materials

Establish clear expectations about how materials will be used and cared for. Consider asking the students to suggest rules for how materials should and should not be used; they are often more attentive to rules and policies that they have helped create.

Initially you may need to place buckets of materials close to students as they work. Gradually, students should be expected to decide what they need and get materials on their own.

Plan a cleanup routine at the end of each class. Making an announcement a few minutes before the end of a work period helps prepare students for the transition that is about to occur. You can then give students several minutes to return materials to their containers and double-check the floor for any stray materials. Most teachers find that establishing routines for using and caring for materials at the beginning of the year is well worth the time and effort.

Encouraging Students to Think, Reason, and Share Ideas

Students need to take an active role in mathematics class. They must do more than get correct answers; they must think critically about their ideas, give reasons for their answers, and communicate their ideas to others. Reflecting on one's thinking and learning is a challenge for all learners, but even the youngest students can begin to engage in this important aspect of mathematics learning.

Teachers can help students develop their thinking and reasoning. By asking "How did you find your answer?" or "How do you know?" you encourage students to explain their thinking. If these questions evoke answers such as "I just knew it" or no response at all, you might reflect back something you observed as they were working, such as, "I noticed that you made two towers of cubes when you were solving this problem." This gives students a concrete example they can use in thinking about and explaining how they found their solutions.

You can also encourage students to record their ideas by building concrete models, drawing pictures, or starting to print numbers and words. Just as we encourage students to draw pictures that tell stories before they are fluent readers and writers, we should help them see that their mathematical ideas can be recorded on paper. When students are called on to share this work with the class, they learn that their mathematical thinking is valued and they develop confidence in their ideas. Because communicating about ideas is central to learning mathematics, it is important to establish the expectation that students will describe their work and their thinking, even in kindergarten.

There is a delicate balance between the value of having students share their thinking and the ability of 5- and 6-year-olds to sit and listen for extended periods of time. In kindergarten classes where we observed the best discussions, talking about mathematical ideas and sharing work from a math activity were as much a part of the classroom culture as sitting together to listen to a story, to talk about a new activity, or to anticipate an upcoming event.

Early in the school year, whole-class discussions are best kept short and focused. For example, after exploring pattern blocks, students might simply share experiences with the new material in a discussion structured almost as list-making:

What did you notice about pattern blocks? Who can tell us something different?

With questions like these, lots of students can participate without one student taking a lot of time.

Later in the year, when students are sharing their strategies for solving problems, you can use questions that allow many students to participate at once by raising their hands. For example:

Luke just shared that he solved the problem by counting out one cube for every person in our classroom. Who else solved the problem the same way Luke did?

In this way, you acknowledge the work of many students without everyone sharing individually.

Sometimes all students should have a chance to share their math work. You might set up a special "sharing shelf" or display area to set out or post student work. By gathering the class around the shelf or display, you can easily discuss the work of every student.

The ability to reflect on one's own thinking and to consider the ideas of others evolves over time, but even young students can begin to understand that an important part of doing mathematics is being able to explain your ideas and give reasons for your answers. In the process, they see that there can be many ways of finding solutions to the same problem. Over the year, your students will become more comfortable thinking about their solution methods, explaining them to others, and listening to their classmates explain theirs.

Games: The Importance of Playing More than Once

Games are used throughout the *Investigations* curriculum as a vehicle for engaging students in important mathematical ideas. The game format is one that most students enjoy, so the potential for repeated experiences with a concept or skill is great. Because most games involve at least one other player, students are likely to learn strategies from each other whether they are playing cooperatively or competitively.

The more times students play a mathematical game, the more opportunities they have to practice important skills and to think and reason mathematically. The first time or two that students play, they focus on learning the rules. Once they have mastered the rules, their real work with the mathematical content begins.

For example, when students play the card game Compare, they practice counting and comparing two quantities up to 10. As they continue to play over days and weeks, they become familiar with the numerals to 10 and the quantities they represent. Later in the year, they build on this knowledge as they play Double Compare, a similar game in which they add and compare quantities up to 12. For many students, repeated experiences with these two games lead them quite naturally to reasoning about numbers and number combinations, and to exploring relationships among number combinations.

Similarly, a number of games in *Pattern Trains and Hopscotch Paths* build and reinforce students' experience with repeating patterns. As students play Make a Train, Break the Train, and Add On, they construct and extend a variety of repeating patterns and are led to consider the idea that linear patterns are constructed of units that repeat over and over again.

Games in the geometry unit, *Making Shapes and Building Blocks*, such as Geoblock Match-Up, Build a Block, and Fill the Hexagons, expose students again and again to the structure of shapes and ways that shapes can be combined to make other shapes.

Students need many opportunities to play mathematical games, not just during math time, but other times as well: in the early morning as students arrive, during indoor recess, or as choices when other work is finished. Games played as homework can be a wonderful way of communicating with parents. Do not feel limited to those times when games are specifically suggested as homework in the curriculum; some teachers send home games even more frequently. One teacher made up "game packs" for loan, placing directions and needed materials in resealable plastic bags, and used these as homework assignments throughout the year. Students often checked out game packs to take home, even on days when homework was not assigned.

Attendance

Taking the daily attendance and talking about who is and who is not in school are familiar activities in many kindergarten classrooms. Through the Attendance routine, students get repeated practice in counting a quantity that is significant to them: the number of people in their class. This is real data that they see, work with, and relate to every day. As they count the boys and girls in their class or the cubes in the attendance stick, they are counting quantities into the 20s. They begin to see the need to develop strategies for counting, including ways to double-check and to organize or keep track of a count.

Counting is an important mathematical idea in the kindergarten curriculum. As students count, they are learning how our number system is constructed, and they are building the knowledge they need to begin to solve numerical problems. They are also developing critical understandings about how numbers are related to each other and how the counting sequence is related to the quantities they are counting.

In *Investigations,* students are introduced to the Attendance routine during the first unit of the kindergarten sequence, *Mathematical Thinking in Kindergarten.* The basic activity is described here, followed by suggested variations for daily use throughout the school year.

The Attendance routine, with its many variations, is a powerful activity for 5- and 6-year-olds and one they never seem to tire of, perhaps because it deals with a topic that is of high interest: themselves and their classmates!

Materials and Preparation

The Attendance routine involves an attendance stick and name cards or "name pins" to be used with a display board. (Many teachers begin the year with name cards and later substitute name pins as a tool for recording the data.)

To make the attendance stick you need interlocking cubes of a single color, one for each class member, and dot stickers to number the cubes.

To make name cards, print each student's first name on a small card (about 2 by 3 inches). Add a photo if possible. If you don't have school photos or camera and film, you might ask students to bring in small photos of themselves from home.

For "name pins," print each student's name on both sides of a clothespin, being sure the name is right side up whether the clip is to the right or to the left.

Name cards might be displayed in two rows on the floor or on a display board. The board should have "Here" and "Not Here" sections, each divided into as many rows or columns as there are students in your class. To display name cards on the board, you might use pockets, cup hooks, or small pieces of Velcro or magnetic tape. Name pins can be clipped down the sides of a sturdy vertical board.

Collecting Attendance Data

How Many Are We? With the whole group, establish the total number of students in the class this year by going around the circle and counting the number of children present.

Encourage students to count aloud with you. The power of the group can often get the class as a whole much further in the counting sequence than many individuals could actually count. While one or two children may be able to count to the total number of students in the class, do not be surprised or concerned if, by the end of your count, you are the lone voice. Students learn the counting sequence and how to count by having many opportunities to count, and to see and hear others counting.

When you have counted those present, acknowledge any absent students and add them to the total number in your class.

Counting Around the Circle Counting Around the Circle is a way to count and double-check the number of students in a group. Designate one person in the circle as the first person and begin counting off. That is, the first person says "1," the second person says "2," and so on around the circle. As students are learning how to count around the circle, you can help by pointing to the person whose turn it is to count. Some students will likely need help with identifying the next number in the counting sequence. Encouraging students to help each other figure out what number might come next establishes a climate of asking for and giving help to others.

Counting Around the Circle takes some time for students to grasp—both the procedure itself and its meaning. For some students, it will not be apparent that the number they say stands for the number of people who have counted thus far. A common response from kindergartners first learning to count off is to relate the number they say to a very familiar number, their age. Expect someone to say, for example, "I'm not 8, I'm 5!" Be prepared to explain that the purpose of counting off is to find out how many students are in the circle, and that the number 8 stands for the people who have been counted so far.

Representing Attendance Data

The Attendance Stick An attendance stick is a concrete model, made from interlocking cubes, that represents the total number of students in the classroom. For young students, part of knowing that there are 25 students in the class is seeing a representation of 25 students. The purpose of this classroom routine is not only to familiarize students with the counting sequence of numbers above 10, but also to help students relate these numbers to the quantities that they represent.

To make an attendance stick, distribute an interlocking cube to each student in the class. After counting the number of students present, turn their attention to the cubes.

We just figured out that there are [25] students in our classroom today. When you came to group meeting this morning, I gave everybody

one cube. Suppose we collected all the cubes and snapped them together. How many cubes do you think we would have?

Collect each student's cube and snap them together into a vertical tower or stick. Encourage students to count with you as you add on cubes. Also add cubes for any absent students.

Ayesha is not here today. Right now our stick has 24 cubes in it because there are 24 students in school today. If we add Ayesha's cube, how many cubes will be in our stick?

Using small dot stickers, number the cubes. Display the attendance stick prominently in the group meeting area and refer to it each time you take attendance.

By counting around, we found that 22 of you are here today. Let's count up to 22 on the attendance stick. Count with me: 1, 2, 3 . . . *[when you reach 22, snap off the remaining cubes].* **So this is how many students are not here—who wants to count them?**

In this way, every day the class sees the attendance stick divided into two parts to represent the students HERE and NOT HERE.

Name Cards or Name Pins Name cards or pins are another concrete way to represent the students. Whereas the attendance stick represents *how many students* are in the class, name cards or pins provide additional data about *who* these people are.

Once students can recognize their name in print, they can simply select their card or pin from the class collection as they enter the classroom each day. At a group meeting, the names can be displayed to show who is here and who is not here, perhaps as a graph on the floor or on some type of display board.

Examining Attendance Data

Comparing Groups In addition to counting, the Attendance routine offers experience with part-whole relationships as students divide the total number into groups, such as PRESENT and ABSENT (HERE and NOT HERE) or GIRLS and BOYS. As they compare these groups, they are beginning to analyze the data and compare quantities: Which is more? Which is less? *How many* more or less? While the numbers for the groups can change on any given day, the sum of the two groups remains the same. Understanding part-whole relationships is a central part of both sound number sense and a facility with numbers.

The attendance stick and the name cards or name pins are useful tools for representing and comparing groups. One day you might use the attendance stick to count and compare how many students are present and absent; another day you might use name cards or pins the same way. Once students are familiar with the routine, you can represent the same data using more than one tool.

To compare groups, choose a day when everyone is in school. Count the number of boys and the number of girls.

Are there more boys than girls? How do you know? How many more?

Have the boys make a line and the girls make a line opposite them. Count the number of students in each line and compare the two lines.

Which has more? How many more?

Use the name cards or the attendance stick to double-check this information.

Once the total number of boys and girls is established, you can use this information to make daily comparisons.

Count the number of girls. Are all the girls HERE today? If not, how many are NOT HERE? How do you know? Can we show this information using the name cards? *[Repeat for the boys.]*

If we know that two girls and two boys are NOT HERE, how many in all are NOT HERE in school today? How do you know? Let's use the name cards to double-check.

When students are very familiar with this routine, with the total number in their class, and with making and comparing groups, you can

pose a more difficult problem. For example:

If we know four students are NOT HERE in school today, how many students are HERE today? What are all the ways we can figure that out, without counting off?

Some students might suggest breaking four cubes off the attendance stick and counting the rest. Others might suggest counting back from the total number of students. Still others might suggest counting up from 4 to the total number of students.

In addition to being real data that students can see and relate to every day, attendance offers manageable numbers to work with. Repetition of this routine over the school year is important; only after students are familiar with the routine will they begin to focus on the numbers involved. Gradually, they will start to make some important connections between counting and comparing quantities.

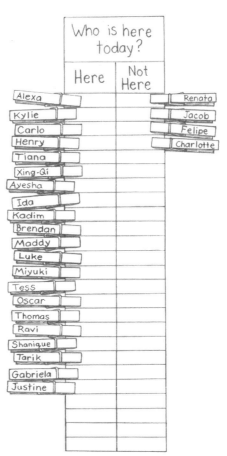

Counting Jar

Counting is the foundation for much of the number work that students do in kindergarten and in the primary grades. Children learn to count by counting and hearing others count. Similarly, they learn about quantity through repeated experiences with organizing and counting sets of objects. The Counting Jar routine offers practice with all of these.

When students count sets of objects in the jar, they are practicing the counting sequence. As in the Attendance routine, they begin to see the need to develop strategies for counting, including ways to double-check and keep track of what they have counted. By recording the number of objects they have counted, students gain experience in representing quantity and conveying mathematical information to others. Creating a new equivalent set gives them not only another opportunity to count, but also a chance to compare the two amounts.

Does my set have the same number as the set in the jar? How do I know?

The jar has 8 and I have 7. I need 1 more because 8 is 1 more than 7.

As students work, they are developing a real sense of both numbers and quantities.

The Counting Jar routine is introduced in the first unit of the kindergarten curriculum, *Mathematical Thinking in Kindergarten.* The basic activity is described here, followed by suggested variations for use throughout the school year on a weekly basis.

Materials and Preparation

Obtain a clear plastic container, at least 6 inches tall and 4–5 inches in diameter. Fill it with a number of interesting objects that are uniform in size and not too small, such as golf or table tennis balls, small blocks or tiles, plastic animals, or walnuts in the shell. The total should be a number that is manageable for most students in your class; initially, 5 to 12 objects would be appropriate quantities.

Prepare a recording sheet on chart paper. At the top, write *The Counting Jar,* followed by the name of the material inside. Along the bottom, write a number line. Some students might use this number line to help them count objects or as a reference for writing numerals. Place each number in a box to clearly distinguish one from another.

Laminate this chart so that students can record their counts on the chart with stick-on notes, write-on/wipe-off markers, or small scraps of paper and tape; these can later be removed and the chart reused.

Also make available one paper plate for each student and sets of countable materials, such as cubes, buttons, keys, teddy bear counters, or color tiles, so that students can create a new set of materials that corresponds to the quantity in the Counting Jar.

Counting

How Many in the Jar? This routine has three basic steps:

- Working individually or in pairs, students count the objects in the Counting Jar.

- Students make a representation that shows how many objects are in the jar and place their response on the chart.

- Students count out another set of objects equivalent to the quantity in the Counting Jar. They place this new set on a paper plate, write their name on the plate, and display their equivalent collection near the Counting Jar.

As you use the Counting Jar throughout the school year, call attention to it in a whole-group meeting whenever you have changed the material or the amount inside the jar. Then leave it in a convenient location for two or three days so that everyone has a chance to count. After most students have counted individually, meet with the whole class and count the contents together.

Note: Some kindergarten teachers use a very similar activity for estimation practice. We exclude the task of estimation from the basic activity because until students have a sense of quantity, a sense of how much 6 is, a sense of what 10 balls look like compared to 10 cubes, it is difficult for them to estimate or predict how large a quantity is. When students are more familiar with the routine and have begun to develop a sense of quantity, you might include the variations suggested for estimation.

One More, One Less When students can count the materials in the jar with a certain amount of accuracy and understanding, try this variation for work with the ideas "one more than" and "one less than." As you offer the Counting Jar activity, ask students to create a set of objects with one more (or less) than the amount in the jar.

Filling the Jar Ourselves When the Counting Jar routine is firmly established, give individuals or pairs of students the responsibility for filling the jar. Discuss with them an appropriate quantity to put in the jar or suggest a target number, and let students decide on suitable objects to put in the jar.

At-Home Counting Jars Suggest to families that they set up a Counting Jar at home. Offer suggestions for different materials and appropriate quantities. Family members can take turns putting sets of objects in the jar for others to count.

Estimation

Is It More Than 5? To introduce the idea of estimation, show students a set of five objects identical to those in the Counting Jar. This gives students a concrete amount for reference to base their estimate on. As they look at the known quantity, ask them to think about whether there are *more than* five objects in the jar. The number in the reference group can grow as the number of objects in the jar changes, and you can begin to ask "Is the amount in the jar more than 8? more than 10?"

More or Less Than Yesterday? You can also encourage students to develop estimation skills when the material in the jar stays the same over several days but the quantity changes. In this situation, students can use reasoning like this:

Last time, when there were 8 blocks in the jar, it was filled up to *here*. Now it's a little higher, so I think there are 10 or 11 blocks.

Calendar

"Calendar," with its many rituals and routines, is a familiar kindergarten activity. Perhaps the most important idea, particularly for young students, is viewing the calendar as a real-world tool that we use to keep track of time and events. As students work with the calendar, they become more familiar with the sequence of days, weeks, and months, and relationships among these periods of time. Time and the passage of time are challenging ideas for most 5- and 6-year-olds, and the ideas need to be linked to their own direct experiences. For example, explaining that an event will occur *after* a child's birthday or *before* a familiar holiday will help place that event in time for them.

The Calendar routine is introduced in the first unit of the kindergarten curriculum, *Mathematical Thinking in Kindergarten.* The basic activity is described here, followed by suggested variations for daily use throughout the school year.

Materials and Preparation

In most kindergarten classrooms, a monthly calendar is displayed where everyone can see it when the class gathers as a whole group. A calendar with date cards that can be removed or rearranged allows for greater flexibility than one without. Teachers make different choices about how to display numbers on this calendar. We recommend displaying all the days, from 1 to 30 or 31, all month long. This way the sequence of numbers and the total number of days are

always visible, thus giving students a sense of the month as a whole.

You can use stick-on labels to highlight special days such as birthdays, class trips or events, non-school days, or holidays. Similarly, find some way to identify *today* on the calendar. Some teachers have a special star or symbol to clip on today's date card, or a special tag, much like a picture frame, that hangs over today's date.

A Sense of Time

The Monthly Calendar When first introducing the calendar, ask students what they notice. They are likely to mention a wide variety of things, including the colors they see on the calendar, pictures, numbers, words, how the calendar is arranged, and any special events they know are in that particular month. If no one brings it up, ask students what calendars are for and how we use them.

At the beginning of each month, involve students in organizing the dates and recording special events on the calendar. The following questions help them understand the calendar as a tool for keeping track of events in time:

> If our trip to the zoo is on the 13th, on which day should we hang the picture of a lion?
>
> Is our trip tomorrow? the next day? this week?
>
> What day of the week will we go to the zoo?

How Much Longer? Many students eagerly anticipate upcoming events or special days. Ask students to figure out how much longer it is until something, or how many days have passed since something happened. For example:

> How many more days is it until Alexa's birthday?
>
> Today is November 4. How many more days is it until November 10?
>
> How many days until the end of the month?
>
> How many days have gone by since our parent breakfast?

Ask students to share their strategies for finding the number of days. Initially many students will count each subsequent day. Later some students may begin to find answers by using their growing knowledge of calendar structure and number relationships:

> I knew there were three more days in this row, and I added them to the three in the next row. That's six more days.

Calculating "how many more days" on the calendar is not an easy task. Quite likely students will not agree on what days to count. Consider the following three good answers, all different, to this teacher's question:

Today is October 4. Ida's birthday is on October 8. How many more days until her birthday?

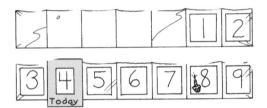

Tess: I think there are four more days because it's 4 . . . [counting on her fingers] 5, 6, 7, 8.

Ravi: There are three more days. See? [He points to the three calendar dates between October 4 and October 8—5, 6, and 7—and counts three date cards.]

Gabriela: It's five more days until her birthday. [Using the calendar, she points to today and counts "1, 2, 3, 4, 5," ending on October 8.]

All of these students made sense of their answers and, considering their reasoning, all three were correct. That's why, when asking "how many more?" questions based on the calendar, it is important also to ask students to explain their thinking.

Numbers on the Calendar

Counting Days The calendar is a place where students can daily visit and become more familiar with the sequence of counting numbers up to 31. Because the numbers on the calendar represent the number of days in a month, the calendar

is actually a way of *counting days*. You can help students with this idea:

Today is September 13. Thirteen days have already gone by in this month. If we start counting on 1, what number do you think we will end up on? Let's try it.

As you involve students in this way, they have another chance to see that numbers represent a quantity, in this case a number of days.

Missing Numbers or Mixed-Up Numbers Once students are familiar with the structure of the calendar and the sequence of numbers, you can play two games that involve removing and rearranging the dates. To play Missing Numbers, choose two or three dates on the monthly calendar and either remove or cover them. As students guess which numbers are missing, encourage them to explain their thinking and reasoning. Do they count from the number 1 or do they count on from another number? Do they know that 13 comes *after* 12 and *before* 14?

Mixed-Up Numbers is played by changing the position of numbers on the calendar so that some are out of order. Students then fix the calendar by pointing out which numbers are out of order.

Patterns on the Calendar

Looking for Patterns Some teachers like to point out patterns on the calendar. The repeating sequence of the days of the week and the months of the year are patterns that help students explore the cyclical nature of time. Many students quickly recognize the sequence of numbers 1 to 30 or 31, and some even recognize another important pattern on the calendar: that the columns increase by 7. However, in order to maintain the focus on the calendar as a tool for keeping track of time, we recommend using the Calendar routine only to note patterns that exist within the structure of the calendar and the sequence of days and numbers. The familiar activity of adding pictures or shapes to form repeating patterns can be better done in another routine, Patterns on the Pocket Chart.

Today's Question

Collecting, representing, and interpreting information are ongoing activities in our daily lives. In today's world, organizing and interpreting data are vital to understanding events and making decisions based on this understanding. Because young students are natural collectors of materials and information, working with data builds on their natural curiosity about the world and people.

Today's Question offers students regular opportunities to collect information, record it on a class chart, and then discuss what it means. While engaged in this data collection and analysis, students are also counting real, meaningful quantities (How many of us have a pet?) and comparing quantities that are significant to them (Are there more girls in our class or more boys?). When working with questions that have only two responses, students explore part-whole relationships as they consider the total number of answers from the class and how that amount is broken into two parts.

Today's Question is introduced in the first unit of the kindergarten curriculum, *Mathematical Thinking in Kindergarten*. The basic routine is described here, followed by variations. Plan to use this routine throughout the school year on a weekly basis, or whenever a suitable and interesting question arises in your classroom.

Materials and Preparation

Prepare a chart for collecting students' responses to Today's Question. If you plan to use this routine frequently, either laminate a chart so that students can respond with wipe-off markers, or set up a blank chart on 11-by-17-inch paper and make multiple photocopies. The drawback of a laminated wipe-off chart is that you cannot save the information collected; with multiple charts, you can look back at data you have collected earlier or compare data from previous questions.

Make a section across the top of the chart, large enough to write the words *Today's Question* followed by the actual question being asked.

Mark the rest of the chart into two equal columns (later, you may want three columns). Leave enough space at the top of each column for the response choices, including words and possibly a sketch as a visual reminder.

Leave the bottom section (the largest part of the chart) blank for students to write their names to indicate their response. Your chart will look something like this:

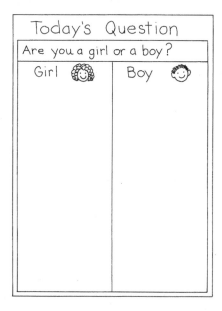

Later in the year, you may want a chart with write-on lines in the bottom section to help students to compare numbers of responses in the two or three categories. Be sure to allow one line for each child in the class. Lines are also helpful guides if you collect data with "name pins," or clothespins marked on both sides with student names, as suggested for the Attendance routine (see illustration on p. 105).

Choosing Questions

Especially during the first half of the school year, try to choose questions with only two responses. With two categories of data, students are more likely to see the part-whole relationship between the number of responses in each category and the total number of students in the class.

As your students become familiar with the routine and with analyzing the data they collect, you may decide to add a third response category. This is useful for questions that might not always elicit a clear yes-or-no response, such as these:

Do you think it will rain? *(yes, no, maybe)*

Do you want to play outside today? *(yes, no, I'm not sure)*

Do you eat lunch at school? *(yes, no, sometimes)*

As you choose questions and set up the charts for this routine, consider the full range of responses and modify or drop the question if there seem to be too many possible answers. Later in the year, as students become familiar with this routine, you may want to involve them in organizing and choosing Today's Question.

Questions About the Class With Today's Question, students can collect information about a group of people and learn more about their classmates. For example:

Are you a boy or a girl?

Are you 5 or 6 years old?

Do you have a younger brother?

Do you have a pet?

Did you bring your lunch to school today?

Do you go to an after-school program?

Do you like ice cream?

Did you walk or ride to school this morning?

Some teachers avoid questions about potentially sensitive issues (Have you lost a tooth? Can you tie your shoes?), while others use this routine to carefully raise some of these issues. Whichever you decide, it is best to avoid questions about material possessions (Does your family have a computer?).

Questions for Daily Decisions When you pose questions that involve students in making decisions about their classroom, they begin to see that they are collecting real data for a purpose. These data collection experiences underscore one of the main reasons for collecting data in the

real world: to help people make decisions. For example:

Which book would you like me to read at story time? (Display two books.)

Would you prefer apples or grapes for snack?

Should we play on the playground or walk to the park today?

Questions for Curriculum Planning Some teachers use this routine to gather information that helps them plan the direction of a new curriculum topic or lesson. For example, you can learn about students' previous experiences and better prepare them before reading a particular story, meeting a special visitor, or going on a field trip, with questions like these:

Have you ever read or heard this story?

Have you ever been to the science museum?

Have you ever heard of George Washington?

For questions of this type, you might want to add a third possible response (*I'm not sure* or *I don't know*).

Discussing the Data

Data collection does not end with the creation of a representation or graph to show everyone's responses. In fact, much of the real work in data analysis begins after the data has been organized and represented. Each time students respond to Today's Question, it is important to discuss the results. Consider the following questions to promote data analysis in classroom discussions:

What do you think this graph is about?

What do you notice about this graph?

What can you tell about [the favorite part of our lunch] by looking at this graph?

If we went to another classroom, collected this same information, and made a graph, do you think that graph would look the same as or different from ours?

Graphs and other visual representations of the data are vehicles for communication. Thinking about what a graph represents or what it is communicating is a part of data analysis that even the youngest students can and should be doing.

Patterns on the Pocket Chart

Mathematics is sometimes called "the science of patterns." We often use the language of mathematics to describe and predict numerical or geometrical regularities. When young students examine patterns, they look for relationships among the pattern elements and explore how that information can be used to predict what comes next. The classroom routine Patterns on the Pocket Chart offers students repeated opportunities to describe, copy, extend, create, and make predictions about repeating patterns. The use of a 10-by-10 pocket chart to investigate patterns of color and shape builds a foundation for the later grades, when this same pocket chart will display the numbers 1 to 100 and students will investigate patterns in the arrangement of numbers.

This routine is introduced in the second unit of the kindergarten curriculum, *Pattern Trains and Hopscotch Paths.* The basic routine is described here, followed by variations for use throughout the school year on a weekly basis.

Materials and Preparation

For this routine you will need a pocket chart, such as the vinyl Hundred Number Wall Chart (with transparent pockets and removable number cards). You will also need 2-inch squares of construction paper of different colors, a set of color tiles (ideally, the colors of the paper squares will match the tiles), and a set of 20–30 What Comes Next? cards. These cards, with a large question mark in the center, are cut slightly larger than 2 inches so they will cover the colored squares. A blackline master for these cards is provided in the unit *Pattern Trains and Hopscotch Paths.* You can easily make your own cards with tagboard and a marking pen.

For the variation Shapes, Shells, and Such, you can use math manipulatives such as pattern blocks and interlocking cubes, picture or shape cards, or collections of small objects, such as buttons, keys, or shells. The only limitation is the size of the pockets on your chart.

What Comes Next?

Before introducing this activity, arrange an a-b repeating pattern in the first row of the pocket chart using ten paper squares in two colors of your choice. Beginning with the fifth position, cover each colored square with a What Comes Next? (question mark) card.

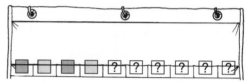

Gather students where the pocket chart is clearly visible and they have a place to work with color tiles, either on the floor or at tables.

Begin by asking students what they notice about the chart. Some may comment on the structure of the chart, some on the two-color pattern, and others may notice the question marks. Explain that each time they see one of these question marks, they should think "What comes next?" and decide what color might be under that card.

Provide each pair with a small cup of color tiles that match the paper squares. Ask students to build the first part of the pattern with color tiles and then predict what color comes next.

Who can predict, or guess, what color is hidden under each question mark on our chart? Use the tiles in your cup to show me what color would come next. How do you know?

Now, with your partner, see if you can make this pattern longer, using the tiles in your cup. Stop when your pattern has ten tiles.

When everyone has made a longer pattern, "read" the pattern together as a whole class. Verbalizing the pattern they are considering often helps students internalize it, recognize any errors in the pattern, and determine what comes next.

This basic activity can be done quickly, especially if students do not build the pattern with tiles. Many teachers integrate this routine into their group meeting time on a regular basis, making one or two patterns on the pocket chart and asking students to predict what comes next.

Initially, use only two colors or two variables in the patterns. In addition to a-b (for example, blue-green) repeating patterns, build two-color patterns such as a-a-b (blue-blue-green), a-b-b (blue-green-green), or a-a-b-b (blue-blue-green-green).

Variations

Making Longer Patterns When students are familiar with the basic activity, they can investigate what happens to an a-b pattern when it "wraps around" and continues to the next line. If the pattern continues in a left-to-right progression, the pattern that emerges is the same one older students see when they investigate the patterns of odd and even numbers on the 100 chart.

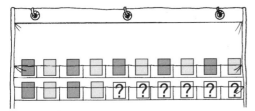

Shapes, Shells, and Such Color is just one variable for patterns; others can be made using a wide variety of materials and pictures.

shell, shell, button, shell, shell, button

triangle, square, triangle, square

♡ ☾ ♡ ☾ ♡ ☾

⇧ ⇩ ⇩ ⇧ ⇩ ⇩ ⇧ ⇩ ⇩

Picture cards, sometimes used by kindergarten teachers to make patterns on the calendar, make great patterns on the pocket chart without the distraction of the calendar elements.

What Comes *Here?* Predicting what comes next is an important idea in learning about patterns. Also important is being able to look ahead and predict what comes *here?* even further down the line. Instead of asking for the *next* color in a pattern sequence, point to a pocket three or four squares along and ask students to predict the

color under that question mark. As you collect responses, ask students to explain how they predicted that color.

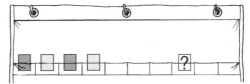

Border Patterns Explore a repeating pattern that extends around the entire outer edge of the pocket chart. Begin by filling the top row of the chart and asking what color would come next if this pattern turned the corner and went down the right side of the chart. Continue adding squares to finish the border. Every few days, begin a new pattern and ask students to help you complete the border. Start with a-b patterns. Gradually vary the pattern type, but continue to use only two colors, trying patterns such as a-a-b, a-b-b, a-a-b-b, or a-a-b-a. Ask students to notice which types make a continuous pattern all around the border and which do not.

With any border pattern, you can include a few What Comes Next? cards and ask students to predict the color of a particular pocket.

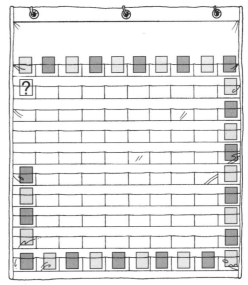

Patterns for Choice Time Hang the pocket chart where students can reach it. During free time or Choice Time, two or three students can work together to make their own pattern on the pocket chart, using colored paper squares or color tiles.

It is likely that more students with limited English proficiency will be enrolled in kindergarten than any other grade. Moreover, many will be at the earliest stages of language acquisition. By correctly identifying a student's current level of English, you can create appropriate stimuli to ensure successful communication when presenting activities from *Investigations*.

The four stages of language acquisition are characterized as follows:

■ **Preproduction** Students do not speak the language at this stage; they are dependent upon modeling, visual aids, and context clues to obtain meaning.

■ **Early production** Students begin to produce isolated words in response to comprehensible questions. Responses are usually *yes, no,* or other single-word answers.

■ **Speech emergence** Students now have a limited vocabulary and can respond in short phrases or sentences. Grammatical errors are common.

■ **Intermediate fluency** Students can engage in conversation, producing full sentences.

You need to be aware of these four levels of proficiency while applying the following tips. The goal is always to ensure that students with limited English proficiency develop the same understandings as their English-speaking peers as they participate in this unit.

Tips for Small-Group Work Whenever possible, pair students with the same linguistic background and encourage them to complete the task in their native language. Students are more likely to have a successful exchange of ideas when they speak the same language. In other situations, teach all students how to make their communications comprehensible. For example, encourage students to point to objects they are discussing.

Tips for Whole-Class Activities To keep whole-group discussions comprehensible, draw simple sketches or diagrams on the board to illustrate key words; point to objects being discussed; use contrasting examples to help explain the attribute under discussion; model all directions; choose students to model activities or act out scenarios.

Tips for Observing the Students Assessment in the kindergarten units is based on your observations of students as they work, either independently or in groups. At times you will intervene by asking questions to help you evaluate a student's understanding. When questioning students, it is crucial not to misinterpret responses that are incomplete simply because of linguistic difficulties.

In many cases, students may understand the mathematical concept being asked about but not be able to articulate their thoughts in English. You need to formulate questions that allow students to respond at their stage of language acquisition in a way that indicates their mathematical understanding.

For example, this unit includes this suggestion for observing the class: "What words do students use to describe their hopscotch paths?" When you are observing students at the speech-emergence and intermediate-fluency stages of acquisition, noting which words they use is appropriate. However, students at earlier stages may not respond adequately simply because they do not have the English skills. Therefore, you need to base your assessment on less-verbal indicators.

With students at the preproduction stage, for example, consider an alternative that calls for a nonverbal response: "Can students show you how they would jump along their hopscotch path?" With students at the early-production stage, you might look for single-word responses: "Can the students tell you how many feet they would use to jump on each part of their path?"

As you observe the students working, keep in mind which guidelines are appropriate for students at the different stages of language acquisition. Following is a categorization of typical questions from this unit.

Questions appropriate for students at the preproduction stage:

- Are students making patterns or are they arranging items in a random way? What type of patterns are they making? Is there variety in their work?

- If students are unable to construct their own pattern, are they able to add on and continue a pattern that you start for them?

- Are students branching out and trying a-b-b or a-a-b patterns?

- What types of materials to students choose? Do they experiment with different materials or stick solely to one or two?

- How easy is it for students to transfer their block pattern to paper? What strategies do they use?

Questions appropriate for students at the early-production and early-speech-emergence stages:

- How do students "read" their pattern? How do they describe the elements in the pattern? by size? color? shape? quantity? position?

- Can students predict what comes next in a pattern?

Questions appropriate for students at the late-speech-emergence and intermediate-fluency stages:

- How do students describe their arrangements? Are they personal? Are they imaginative? Are they descriptive?

- How accurate are their statements? Are descriptive words (size, color, shape, quantity) used correctly?

- What are students saying about the similarities and differences among arrangements of color?

- Can students explain their reasoning, logic, or strategies for determining what is missing?

The following activities will help ensure that this unit is comprehensible to students who are acquiring English as a second language. The suggested approach is based on *The Natural Approach: Language Acquisition in the Classroom* by Stephen D. Krashen and Tracy D. Terrell (Alemany Press, 1983). The intent is for second-language learners to acquire new vocabulary in an active, meaningful context.

Note that *acquiring* a word is different from *learning* a word. Depending on their level of proficiency, students may be able to comprehend a word upon hearing it during an investigation, without being able to say it. Other students may be able to use the word orally, but not read or write it. The goal is to help students naturally acquire targeted vocabulary at their present level of proficiency.

missing, add, same, different

1. Draw two simple flowers side by side, identical except that one has no leaves.

2. With the group, compare the two flowers, pointing to each part as you ask about it.

   Do both flowers have the same number of petals? Do both flowers have a stem? Do both flowers have leaves?

3. Explain that the two flowers are *different* because one *[point]* is *missing* its leaves. Ask what you could do to make both the *same*.

4. As you draw the missing leaves, explain that you are *adding* leaves.

5. Ask if they can think of other things you could add to the drawings. If they cannot verbalize their ideas, let them draw their own additions. Describe what they add.

   Sione added a sun. Mika added grass.

6. Without erasing the flowers, draw two more almost identical items, such as cars, houses, trees, or children, and continue with questions in the same format.

7. Refer back to each drawing. Challenge students to recall what was originally *missing* and what was *added* on.

color names (red, white, yellow, black, blue, orange, green, brown, purple), next

1. Prepare squares of construction paper in all different colors and lay them in a row. Identify the color of each square as you hop a teddy bear counter from one to the next.

   The *next* square is red. The *next* one is purple.

2. Provide teddy bear counters to the students. As you call out a color name, students hop their teddy bear to that square of paper and repeat the color name.

squares, corner, edge

1. Prepare a 6-inch square of construction paper and a set of at least 20 color tiles or 1-inch squares of paper. Display these and identify them all as *squares*.

2. Identify each *edge* of the large square as you trace it with your finger. Also identify the four *corners* of the large square.

3. Distribute tiles or paper squares to the students. Call on volunteers to place these along the inside edge of the large square.

   Who can put a small square on one edge of this large square?

   Who can put a small square at a corner of this large square?

4. Continue until students have made a border all around the edges of the larger square.

up, down

1. Set up blocks or books in the shape of a simple staircase. Hop a teddy bear counter up and down the stairs, using the words *up* and *down* as you move the bear.

2. Place several teddy bear counters at the top and bottom of the stairs. Ask students, one at a time, to move a bear either *up* or *down* the stairs.

3. Choose one student to be the caller and tell the others how to move bears by calling out *"Up"* or *"Down."*

Blackline Masters

_____ , 19 _____

Dear Family,

Our class is beginning a mathematics unit called *Pattern Trains and Hopscotch Paths*. For the next few weeks, we will explore questions like these: What makes a pattern a pattern? How do patterns give us information so that we can predict what comes next?

Being able to recognize patterns is an important tool in mathematics. The children will have many opportunities to copy, create, and extend patterns using materials such as pattern blocks, color tiles, and interlocking cubes.

You can help your child at home by talking about patterns.

- Look for patterns in the environment. Where do you see patterns? How are patterns made? How do they use shape, color, size, position, or quantity? Can you find patterns in the music you hear or in the stories you read or tell?

- Look at the clothing in your child's closet. Which items have patterns and which do not? Your child may want to sort his or her clothes into two groups: those with patterns and those without.

- Make patterns together. Lots of household items are fun to make patterns with: buttons, caps and bottle tops, coins, and keys are just a few. You can also take turns adding on to each other's pattern.

- Try physical pattern routines with motions, such as clapping your hands and tapping your knees in a repetitive pattern. For example: *clap, clap, tap; clap, clap, tap; clap, clap, tap.* Start a pattern and see if your child can predict what might come next. Then reverse the game, with your child making a pattern for you to extend.

Have fun exploring these ideas together.

Sincerely,

PATTERN PATHS

Copy on card stock to make one Pattern Path for each student. Cut along the outside lines and glue strips end to end. If possible, laminate the paths.

WHAT COMES NEXT? CARDS

Copy on heavy card stock, 2–3 sheets per class. If you do not have card stock, mount cards on a dark background to prevent see-through. Laminate if possible.

?	?	?
?	?	?
?	?	?
?	?	?

COLOR CARS

Copy on card stock, 1 sheet for each Make-a-Train bag.
Cut apart and color in different combinations.

ADD TO OUR PATTERN MUSEUM

Dear Family,

The children are very excited by our investigations with patterns. They are eager to display and share some of their work. We are currently setting up a special area in our classroom called The Pattern Museum. Along with the patterns that we are making in class, we would like to display examples of real objects with repeating patterns on them.

Please help your child find one or more examples to put on display in our Pattern Museum. You might find repeating patterns in pieces of wallpaper, fabric, or flooring. There may be patterns on your dishtowels or a pattern on a picture frame. Encourage your child to look through clothing for something like a shirt with a repeating pattern.

Put your name on any item you send in. All items will be returned when our project is over. We will be careful with items from home. However, please do not send in anything breakable, valuable, or precious to you.

You are welcome to come in and visit our Pattern Museum. If you are interested, please let me know. We can arrange a time before, during, or after school.

Thanks for your continued support of the work we are doing in our kindergarten classroom.

Border Mat A

Border Mat B

Border Mat C

12 chips in All

Pattern													Count		
													Red	Yellow	Total
○	○	○	○	○	○	○	○	○	○	○	○				
○	○	○	○	○	○	○	○	○	○	○	○				
○	○	○	○	○	○	○	○	○	○	○	○				
○	○	○	○	○	○	○	○	○	○	○	○				

STAIRCASE CARD A

Copy on heavy card stock.

116

ONE-INCH GRID PAPER

Choice Board art for
What's Missing?

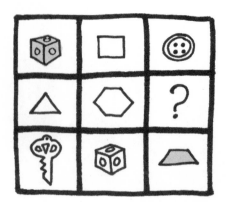

Choice Board art for
Making Patterns

Choice Board art for
What Comes Next?

Choice Board art for
Pattern Block Snakes

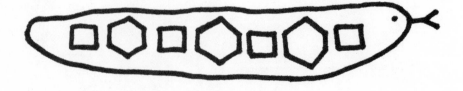

Choice Board art for
Break the Train

Choice Board art for
Make a Train

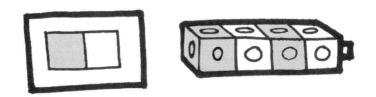

Choice Board art for
Add On

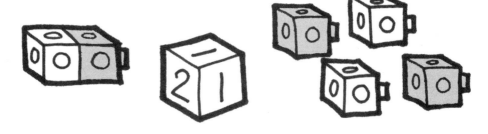

Choice Board art for
Hopscotch Paths

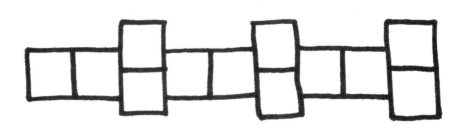

Choice Board art for
Tile Paths

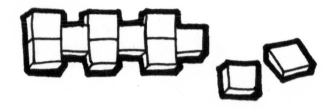

Choice Board art for
Color Tile Borders

Choice Board art for
12 Chips

Choice Board art for
Staircase Pattern

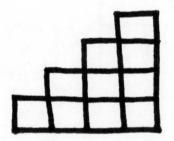

PATTERN BLOCK CUTOUTS (page 1 of 6)

Duplicate these hexagons on yellow paper and cut apart.

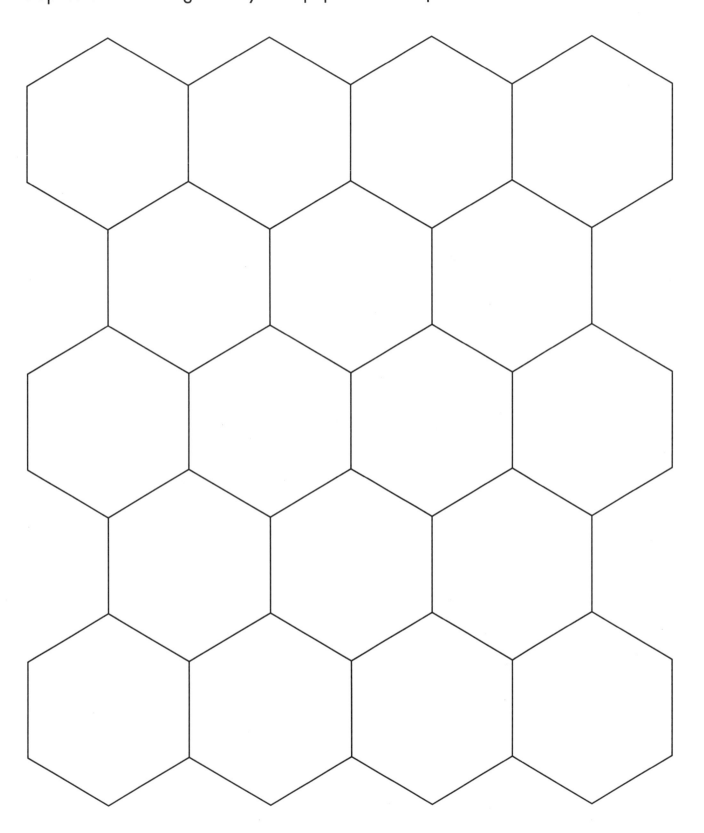

PATTERN BLOCK CUTOUTS (page 2 of 6)

Duplicate these trapezoids on red paper and cut apart.

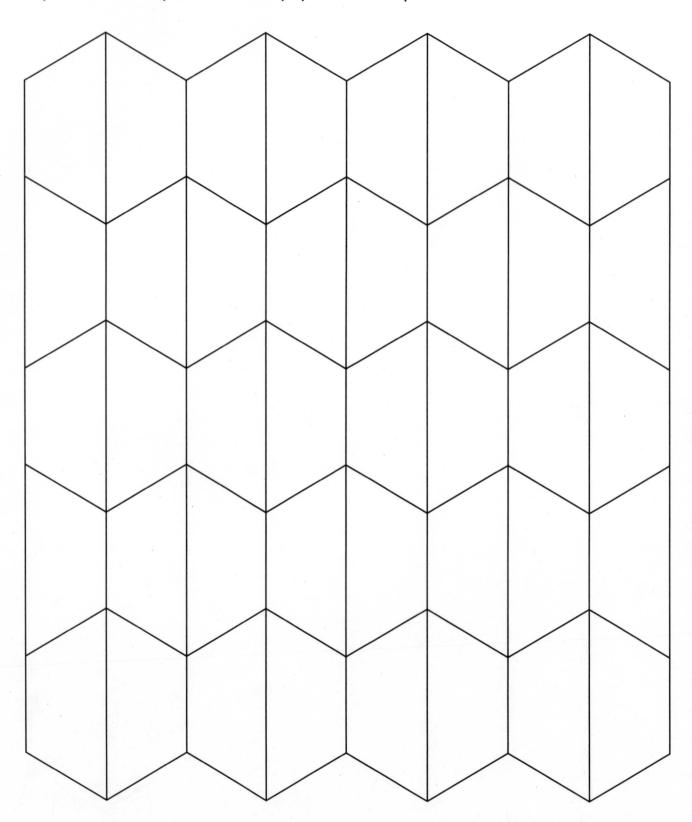

Duplicate these triangles on green paper and cut apart.

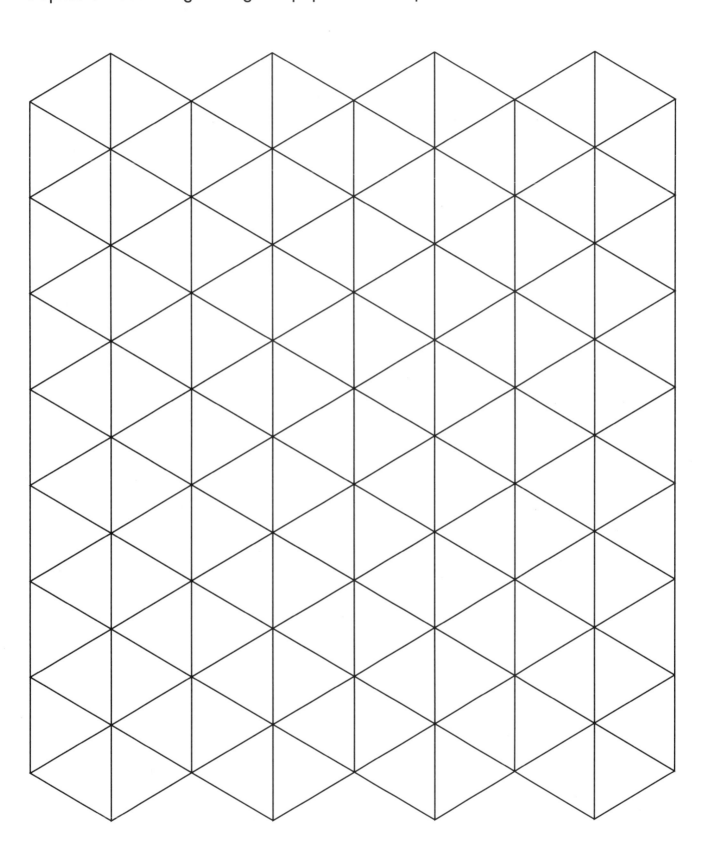

Duplicate these squares on orange paper and cut apart.

PATTERN BLOCK CUTOUTS (page 5 of 6)

Duplicate these rhombuses on blue paper and cut apart.

Duplicate these rhombuses on tan paper and cut apart.